Modern Antenna

The modern antenna book series mainly covers the related antenna theories and technologies proposed and studied in recent years to solve the bottleneck problems faced by antennas, including binary coded antenna optimization method, artificial surface plasmon antenna, complex mirror current equivalent principle and low profile antenna, generalized pattern product principle and generalized antenna array, cross dielectric transmission antenna, metamaterial antenna, as well as new antenna technology and development. This series not only presents the important progress of modern antenna technology from different aspects, but also describes new theoretical methods, which can be used in modern and future wireless communication, radar detection, internet of things, wireless sensor networks and other systems.

The purpose of the modern antenna book series is to introduce new antenna concepts, new antenna theories, new antenna technologies and methods in recent years to antenna researchers and engineers for their study and reference. Each book in this series is thematic. It gives a comprehensive overview of the research methods and applications of a certain type of antenna, and specifically expounds the latest research progress and design methods. As a collection, the series provides valuable resources to a wide audience in academia, the engineering research community, industry and anyone else who are looking to expand their knowledge of antenna methods. In addition, modern antenna series is also open. More antenna researchers are welcome to publish their new research results in this series.

More information about this series at https://link.springer.com/bookseries/16786

Junping Geng · Ronghong Jin

Antenna Optimization and Design Based on Binary Coding

Springer

Junping Geng
Department of Electronic Engineering
Shanghai Jiao Tong University
Shanghai, China

Ronghong Jin
Department of Electronic Engineering
Shanghai Jiao Tong University
Shanghai, China

Modern Antenna
ISBN 978-981-16-7964-3 ISBN 978-981-16-7965-0 (eBook)
https://doi.org/10.1007/978-981-16-7965-0

This Springer imprint is published by the registered company Springer Nature Singapore Pte Ltd.
The registered company address is: 152 Beach Road, #21-01/04 Gateway East, Singapore 189721, Singapore

Preface

With the rapid development of remote sensing, wireless transmission, and radar technique, antenna has the bottleneck of the wireless system. There are ten thousands of researchers and engineers are studying and design antenna. But they mostly design the antenna from their experiences or their original model from the books or others papers, and then revise the structure or parameters. So, what is the optimum antenna for your design task? It is very difficult to answer. As we said, the design process is disturbed by your prior knowledge and experiences.

In fact, it is impossible to completely abandon our prior knowledge and experience to design antennas. In order to decrease these disturbances, optimization methods are introduced to assist engineers to rapidly design antennas with good performance and to solve the inverse scattering problem of the antenna by constructing the system with rules, aims, and basic units. At present, the optimization methods applied in antenna design mainly include particle swarm optimization (PSO) algorithm, genetic algorithm (GA), and simulated annealing algorithm. Based on the long research experience and results of our group, this book focuses on the application of optimization algorithms in multi-frequency antenna, low-profile antenna, ultra-wideband antenna, and so on.

In Chap. 1, we introduce background and the importance of antenna optimization. Then, the development and trend of antenna optimization based on AI are discussed. In Chap. 2, we emphatically introduced the basic concepts and principles of particle swarm optimization algorithms and improved particle swarm optimization algorithms, as well as the flow of their use, followed by a brief description of other optimization algorithms such as GA.

Then in Chaps. 3–9, examples of using optimization algorithms to optimize antenna are explained in detail. All the results mentioned in this book are published journal, articles, and patents by our groups. In Chap. 3, a multi-frequency antenna is successfully proposed by the steps combining rough designs and precise designs. The condition that the new grids can be only placed near the old one is set to avoid the discontinuity. A-CLPSO is introduced to avoid premature convergence. In Chap. 4, carved patch antenna part and carved middle ground based on particle swarm optimization (PSO) method are proposed, which can work well on the metal ground at

Chinese RFID frequency band—840–845 MHz. In Chap. 5, two parasitic patches are established by grid and optimized by PSO. Placed under the spiral antenna, the antenna requires more broad bandwidth, stable radiation pattern, and better circular performance. In Chap. 6, a binary-code mm-wave antenna with broadband, dual-polarization, and wide beamwidth is proposed. The driven patch and the parasitic patch are formed by series of small rectangular units optimized by binary codes with multi-objects. The designed mm-wave antenna element can work from 21.8 to 33.2 GHz. In Chap. 7, we first use the genetic algorithm to optimize the parameters of the UWB antenna. In Chap. 8, discrete hexagon grids in parasitic layer are presented, which guarantee the line-to-line connection between adjacent elements. After optimizing the hexagon grids, a low-profile antenna with wide CP bandwidth and stable unidirectional pattern is obtained. In Chap. 9, the configuration of the particle swarm optimization algorithm is presented. And a continuous and smooth structure generating method is described. In the end, the optimized result and experiment data are analyzed. The book is not only intended for researchers investigating high performance antenna, and antenna design engineers working on new antenna design and applications, but also being textbooks for undergraduate and graduate students who are interested in antenna technology.

Shanghai, China

Junping Geng
Ronghong Jin

Acknowledgements The author would like to thank Xiang Liu, Xiaonan Zhao, Lei Wang, Hao Wu, Xiaohui Tao, Bangda Zhou, Min Ding, and Luyang Duan for their early works. The author would also like to thank Haobo Wu, Jingzheng Lu, Da Su, Yangzhou Zhang, Chaofan Ren, Erwei Liu, Kun Wang, Han Zhou, Xuxu Cheng, Jiawei Han, Weinan Gao, Jing Zhang, and Silei Yang for their support and assistance during the manuscript preparing.

Contents

About the Authors

Dr. Junping Geng received his B.E degree in plastic working of metals, M.S. degree in corrosion and protection of equipment, and Ph.D. degree in circuit and system from the Northwestern Polytechnic University, Xian, China, in 1996, 1999, and 2003, respectively.

From 2003 to 2005, he was a Postdoctoral Researcher with Shanghai Jiao Tong University, Shanghai, China. In April 2005, he joined the faculty of the Electronic Engineering Department, Shanghai Jiao Tong University. He was promoted to an associated professor in 2008, and he is now a doctoral supervisor. From 2010 to 2011, he was a visiting scholar with the Institute Electrical and Computer Engineering, University of Arizona, AZ, USA.

Besides, he is among the editorial board of *Nano Material, International Journal of Antenna and Propagation and International Journal of Aerospace Engineering*. He is also a committee member of Antenna Society of Chinese Institute of Electronics. He has been the Senior member of IEEE in 2017, and he also is the editor-in-chief of the *Modern Antenna* book series (Springer). He has served as member of Technology Program Committee (TPC) and session chairs for more than 30 international conferences.

He is mainly engaged in the teaching and research work on the fields covering antenna and electromagnetic theory, array signal processing, wireless communications, and nano technology, etc.

In 2015, he received the best paper award in IEEE MAPE 2015. In 2013, he was awarded the third rank of Shanghai Award for Natural Sciences. In 2010, he received the best paper award in IEEE iWAT2010. In 2008, he won the second prize of Chinese National Technology Innovation Awards as the third co-author. In 2007, he won the first prize of Technology Innovation Awards of the Ministry of Education as the fourth co-author. In 2009, he won the second Prize of "Hengshan Liangci" Excellent Paper. He won the Shanghai Jiao Tong University excellent annual assessment in 2007, 2014, 2016, 2017, and 2018 respectively, and he also won the third prize for excellent teachers in 2008, excellent in the appointment period 2014–2016 in Shanghai Jiao Tong University too. He won the third prize of postdoctoral award fund in 2006.

So far he had published over 380 papers at home and abroad, about 127 of them in International Journals. He has applied over 150 invention patents, and 100 of them have been granted.

In addition, he has published three monographs of *Omnidirectional Slots Antenna, Spoof Surface Plasmon Polarizations Antenna, Smart Antenna in Wireless Communication*, and one textbook of *Introduction to Computational Electromagnetism*, and provided chapters for three books.

He has been in charge of or involved in over 30 projects including Chinese National Fund, the State Key Program of National Natural Science, 973, Innovation group and Shanghai Research Projects.

Dr. Ronghong Jin received his B.E, M.E and Ph.D. degrees in 1983, 1986 and 1993 respectively all from Department of Electronic Engineering of Shanghai Jiao Tong University. He was promoted to a professor with exceptive admission in 1996. He then stayed in Tokyo Institute of Technology of Japan from 1997 to 1999 as a postdoctoral visiting scholar. From 2001 to 2002, he was invited to work as an honorary professor rank research fellow in TAO and special research fellow in CRL in Japan. Later in 2006, he was invited as guest professor in the School of Information in University of Wollongong of Australia and in 2010 he was appointed as distinguished visiting scientist by Information, Communication and Technology (ICT) Center of Australian Commonwealth Scientific and Industrial Research Organization (CSIRO). From 2005 to 2011 he served as chair of Shanghai Session of IEICE. At present, he is IEEE Fellow, a Distinguished professor of SEIEE in Shanghai Jiao Tong University. Besides, he is among the editorial board of Chinese Journal of Radio Science. He is also a hournor member of Antenna Society and a member of Radio Wave Society of Chinese Institute of Electronics.

He is mainly engaged in the teaching and research work on the fields covering antenna and electromagnetic theory, array signal processing, and wireless communications, etc. In 2018, he was entitled the Outstanding Ph.D. Thesis Mentor of Chinese Institute of Communication. In 2012, he was entitled the Shanghai Outstanding Ph.D. Thesis Mentor and won the nomination of its Chinese National Award. In 2010, he received the best paper award in IEEE iWAT2010. He won the second prize of Chinese National Technology Innovation Awards as the second co-author, the first prize of Technology Innovation Awards of the Ministry of Education as the second co-author, the second prize of Chinese National Award for Natural Sciences as the fifth co-author and the first prize nomination by Natural Science Awards of the Ministry of Education as the fourth co-author in 2008, 2007, 2004 and 2003 respectively. In addition, in 2000, he was sponsored by the Ministry of Education for the Excellent Young Teachers' Program. In 2013, he was awarded the third rank of Shanghai Natural Sciences Awards as the first author. In 1999, he was awarded the third rank of Shanghai Science and Technology Progress Awards. He was also the leader of a Shanghai elaborate course:<<Electromagnetic Fields and Waves>>.

So far he had published over 530 papers at home and abroad, about 170 of them in International Journals. He has applied over 150 invention patents, and more than 100 of them have been granted. Also, he is also the co-author of two text books and four academic books.

He has been in charge of or involved in over 90 projects including Chinese National Fund, the State Key Program of National Natural Science, 973, Innovation group and Shanghai Key Research Projects.

Abbreviations

2D	Two dimensions
3D	Three dimensions
A-CLPSO	Adaptive-comprehensive learning particle swarm optimization
ACO	Ant colony optimization
ACP	Automatic cell planning
AI	Artificial intelligence
BA	Bat algorithm
BP	Back propagation
CLPSO	Comprehensive learning particle swarm optimization
CMOS	Complementary metal–oxide–semiconductor
CP	Circular polarization
CSA	Cuckoo search algorithm
CST	Commercial electromagnetic software
EBG	Electronic band gap structure
EM	Electromagnetic
EP	Evolutionary planning
ES	Evolution strategy
FCC	Federal Communications Commission
FDTD	Finite-difference time domain
FSA	Fish school optimization
GA	Genetic algorithm
HBA	Honey bee algorithm
IFF	Identification, friend or foe
MDT	Minimization of drive test
MEC	Moving edge calculation
MR	Measurement report
PBG	Photonic band gap structure
PSO	Particle swarm optimization
RFID	Radio-frequency identification
SA	Simulated degradation algorithm
TD-SCDMA	Time division-synchronous code division multiple Address

TSA	Tabu search algorithm
UWB	Ultra-wideband
WLAN	Wireless local area network

List of Figures

List of Tables

Chapter 1
Introduction

1.1 Background and Demand

The antenna is the key component of the communication, radar, navigation, radio, television, and other radio system. The antenna is just like ears and eyes of the whole communication system, which is the interface between the system and external media. All systems based on radio wave communication are inseparable from the antenna, and for these communication systems, the selection and design of the antenna will directly affect performance of the whole system.

Designing a proper antenna with excellent performance will greatly reduce the design pressure and cost of the whole communication system back-end and improve the communication performance of the communication system. Since Marconi first designed the antenna to realize long-distance wireless communication in 1890s, people have done a lot of research and design on the antenna. According to different application fields and scenarios, people have designed various types of antenna.

Antenna optimization is very important to antenna design, but more in parameters or size optimization, and less in geometry design optimization.

It is very convenient to digitize the size parameters, but it is very difficult to digitize the structure. When the size parameters and geometry structures are in one system, just like the analog parameters and digital discrete digital parameters are mixed, we need a uniform expression.

Antenna design is a multi-aim problem, which includes frequencyband, size, pattern, polarization and gain. So, the adaptive weighted fitness function based on the appropriate optimization method is very necessary. must be adaptive. Uniform array can increase the antenna performances. However, the power of the array is not limited to linear weight feeding, and the optimized nonlinear weight change can better tap the potential of the array.

J. Geng and R. Jin, *Antenna Optimization and Design Based on Binary Coding*, Modern Antenna, https://doi.org/10.1007/978-981-16-7965-0_1

1.2 Research Development of the Optimum Antenna Design

1.2.1 Antenna Optimization

Antennas design is an anti-process of EM problem. It usually comes from the field demands and boundary conditions, and then to find the optimized current distribution in the source region, that is the optimized antenna. Certainly, this process usually requires human intervention. The original basic model of the antenna and the human interventions are usually derived from the experience of the designer.

In 1975, Professor J. Holland of the Michigan University proposed genetic algorithm [1], which borrowed from natural selection and natural genetic mechanism of the biological world to solve extreme value problem. It is a bionic stochastic optimization algorithm. This algorithm is widely used and can be used to solve complex and nonlinear problems that traditional search methods cannot solve. Many scholars have designed different codes to express feasible solutions to different problems in different fields, but they all imitate the selection, crossover and mutation mechanisms of the biological world. In 1994, scholars from the University of California proposed to use genetic algorithms to optimize antenna array, mainly for one-dimensional and two-dimensional antenna array pattern optimization [2]. In 1997, they systematically summarized the application of genetic algorithm in electromagnetic field, showing the broad application prospect of genetic algorithm. Ding Min combined genetic algorithm and finite difference time domain method for parameter optimization and designed a small ultra-wideband antenna [3]. Later, it combined two-dimensional genetic algorithm with finite difference time domain method to realize automated design of ultra-wideband antennas and band-stop ultra-wideband antennas [4, 5].

In 1995, at the IEEE International Conference on Neural Networks, J. Kennedy and R. C. Eberhatr proposed particle swarm optimization algorithm. This algorithm originated from the study of bird predation behavior. This algorithm is similar to genetic algorithm, they both belong to evolutionary algorithm [6]. The system is initialized as a set of random solutions, and then the optimal solution is obtained through iteration. In each iteration, the particle updates itself by tracking the optimal solution obtained by the particle itself and the current optimal solution of the group. The algorithm is simpler than genetic algorithm rules without crossover and mutation operations, and at the same time it has the advantages of high accuracy and fast convergence, which shows its superiority in practical applications. Particle swarm algorithm is also often used to design array antenna systems, such as the comprehensive optimization of pattern [7]. Professor Jin Ronghong from Shanghai Jiaotong University proposed an improved PSO algorithm and applied it to pattern synthesis of linear arrays, verifying that the improved algorithm has a good optimization effect [8]. Wu Hao improved particle swarm algorithm, proposing A-CLPSO algorithm, and designed a small multi-band monopole antenna which coverages TD-SCDMA (1.88–2.05 GHz), IEEE 802.11b/g (2.4–2.484 GHz), IEEE 8.02.11a (5.15–5.85 GHz) and Wimax (5.15–5.85 GHz) [9]. On this basis, scholars in literature [10] used Wu

Hao's optimization method to design a broadband microstrip antenna with directional radiation in 0.82–2.1 GHz frequency band by opening rectangular holes in metal parasitic patch. In literature [11], scholars designed a microstrip antenna suitable for RFID frequency band through this method. But Wu Hao's optimization algorithm has poor versatility, which also limits its wide application. In literature [12], the author uses PSO algorithm to optimize a patch antenna. He used PSO algorithm optimizes the shape of patch and keeped the edge of patch as a smooth curve. Although the experience of the antenna designer is not very important in this method, it also means that many traditional antenna forms with better performance cannot be used and played well.

Nowadays, many commercial electromagnetic simulation software also have the function of antenna optimization, and most of them also use genetic algorithm, particle swarm algorithm, Newton method and other optimization methods. Although it has versatility, it mainly optimizes the parameters of the built antenna model, which means that the antenna designer needs to have a wealth of antenna experience.

1.2.2 Antenna Optimization Based on AI

1.2.2.1 Research Status of Artificial Intelligence Algorithms

From the 1960s, people have tried to carry out a series of simulations of different biological intelligences in nature to find new intelligent algorithms from the natural rich information. For the limitations of traditional optimization algorithms, it is very difficult to solve the more complex optimization problems. However, the intelligent algorithm based on biological simulation does not depend on the mathematical characteristics of the optimization problem itself, nor does it require that the function studied to be continuous and differentiable. It processes and calculates input information directly in the data layer. Intelligent algorithm is very suitable for solving the problem that it is difficult to establish a formal model. We call these algorithms discovered and inspired from nature artificial as intelligence algorithms, which is also the crystallization of people's research for more than half a century. Most AI algorithms are uncertain about the search direction of the objective function. The next action of the algorithm is completely determined by the probability, so the artificial intelligence algorithm belongs to the probability algorithm. Probabilistic algorithms include evolutionary algorithm, evolutionary strategy [13], evolutionary programming [14], genetic algorithm [15], etc. Other probabilistic algorithms based on swarm intelligence include ant colony optimization algorithm [16], particle swarm optimization algorithm [17], fish swarm optimization algorithm [18]. Other common probabilistic algorithms include simulated annealing algorithm [19], tabu search algorithm [20, 21], etc. the probabilistic algorithms are shown in Table 1.1.

Table 1.1 The development history of some artificial intelligence algorithms

Year of proposal	Artificial intelligence algorithm name	English abbreviations	Proposer
1963	Evolution strategy	ES	Rechenberg. Sehwefel
1966	Evolutionary planning	EP	Fogel
1975	Genetic algorithm	GA	Holland
1983	Simulated degradation algorithm	SA	Kirkpatrick, Gelatt, Vecchi
1989	Tabu search algorithm	TSA	Glover
1991	Ant Colony Optimization	ACO	Colomi, Dorigo, Maniezzo
1995	Particle swarm optimization	PSO	Kennedy, Eberhart
2002	Fish school optimization	FSA	Xiao-leiLi
2005	Bee colony algorithm	HBA	Karaboga
2009	Cuckoo search algorithm	CSA	Yang, Deb
2010	Bat Algorithm	BA	Yang

1.2.2.2 Self-Tuning System of Antenna Feeder Based on Al Algorithm

With the popularity of mobile Internet services, people put forward higher and higher requirements for the quality of wireless mobile network. The performance of antenna feed system directly affects the performance and quality of wireless mobile network, so the optimization of antenna feed system is particularly important. Traditional antenna feed optimization depends on the experience of optimization personnel, which has the problems of large resource investment, inaccurate data parameters and low optimization efficiency. Recently, in order to solve many limitations of traditional antenna and feed optimization methods, some intelligent antenna and feed optimization methods have been proposed, such as antenna and feed intelligent optimization method based on minimum drive test [22] and optimization method based on MEC and K-means clustering [23]. Among them, the intelligent optimization method of antenna and feeder based on MDT requires mobile terminals supporting MDT to report the measurement report information including longitude and latitude, but not all mobile terminals support MDT function; the optimization method based on MEC and K-means clustering is adopted for large-scale data sets. The convergence speed is slow and it is easy to fall into local minimum. Reference [24] uses MDT and MR to collect users' wireless quality data and longitude and latitude information, and then uses density based on clustering method and adaptive clustering algorithm to cluster the collected wireless coverage index and longitude and latitude data. ACP (automatic cell planning) is an intelligent automatic optimization technology based on industrial parameters, electronic map, antenna pattern, load, Mr or MDT data sources, aiming at overlapping coverage and downlink rate. Network coverage, capacity and quality issues to achieve the best overall network performance [25]. With the rapid increase of the number of base stations, how to complete the intelligent optimization of antenna feeder with high efficiency and quality has become an urgent problem.

1.2.2.3 Research on 5G Large-Scale Antenna Pattern Configuration Method Based on Artificial Intelligence

Network artificial intelligence refers to the application of artificial intelligence technology to the operator's network to replace or optimize the work currently performed by humans through the intelligent or intelligent subsystem of the network, so that the operator can more conveniently and efficiently provide better network services. 5G Massive MIMO faces the challenges of diversified coverage scenarios and more complicated parameter configuration during deployment. It is necessary to combine artificial intelligence algorithms to give full play to the performance advantages of Massive MIMO, improve network optimization efficiency, and save network operation and maintenance costs. In this paper, starting from the common use of synchronization and broadcast channels in 5G scenarios, the scene fingerprint is constructed through beam-level user distribution, an antenna broadcast mode configuration model is proposed, and the antenna broadcast weight adaptive configuration optimal solution is obtained through multiple iterations of the ant colony algorithm. Clearly configure the basic process and specific use cases. Adaptive optimization of 5G Massive MIMO antenna weights through AI can not only replace part of manual network planning and network optimization, but also reduce the weight search cycle based on big data analysis, and dynamically track changes in UE distribution and inter-cell interference to ensure timely coverage of users in the cell Maximum number, considering the volume of traffic. As the main body of information infrastructure construction and operation, operators have obvious advantages in developing artificial intelligence technology and expanding applications: the number of users is huge, they are at the forefront of information networks, and they gather the entire network data carried by the Internet, mobile Internet, and dedicated lines. And part of the local area network data traffic can obtain user core data; sufficient accumulation in information and data can provide the basic guarantee for big data storage and processing for the development of artificial intelligence. Artificial intelligence will be widely and deeply applied in the telecommunications field.

1.3 Developing Trend and Problem Faced

1.3.1 The Developing Trend to Antenna Optimization

Traditional antenna design methods usually rely on the work experience and theoretical knowledge of antenna designers. The design process is not only complicated but also time-consuming, and it is also difficult to achieve optimal design. In recent years, antenna optimization design has been paid attention and researched. It uses the antenna numerical calculation method to perform full-wave numerical simulation of the antenna related performance and uses modern optimization algorithms such as genetic algorithm to realize the computer-aided design of the antenna structure. Its basic principle is to transform the antenna design into a genetic algorithm search and optimization process. Existing studies have shown that [1, 26]. antenna optimization design can save designers a lot of energy, while broadening the antenna design range, improving design accuracy, has become a new hot spot in modern antenna research. However, in the optimization design process, it is necessary to repeatedly perform the antenna full-wave numerical simulation, which occupies most of the time-consuming optimization design. The neural network algorithm can quickly perform a large number of calculations and can fully approximate arbitrarily complex nonlinear relationships, so it is very suitable for solving problems such as antennas that have complex nonlinear relationships between structural parameters and related performance. Literature [27] uses a 3-layer BP neural network model to predict the S11 of a patch antenna, which greatly improves the antenna design efficiency. Literature [28] proposed a method based on radial basis neural network and used to calculate the pattern of uniform linear array composed of collinear short dipoles and parallel short dipoles. This method is also suitable for short dipoles. Calculation of sub-plane array [29]. However, they are all purely applying the neural network to the prediction of antenna-related performance, and the prediction is completely out of the full-wave numerical simulation. This may cause the prediction error of the neural network algorithm to adversely affect the result of the antenna optimization design.

1.3.2 The Problem Faced in the Developing

First of all, the current antenna optimization design methods require artificial selection of prior models, and human intervention is needed in the middle.

Secondly, the weight distribution of multi-objective optimization needs to be modified and adjusted artificially in the middle process.

Thirdly, many times, it is difficult to give the range of optimization parameters accurately, which affects the efficiency of optimization.

Fourth, the optimization step is difficult to give, it is very likely that the optimization is too rough, it may also waste time.

Fifthly, it is easy to enter the dead cycle or local optimum.

Sixth, there are too many parameters and too many intermediate process results to choose.

Seventh, it is difficult to form multi-objective evaluation criteria accurately, which leads to lengthy calculation and missing good results.

Eighth, the coarse model (discretization) optimization based on trellis coding and the fine model (continuity) optimization based on fine-tuning cannot be completed adaptively.

1.4 Content and Construction of the Book

In first chapter, we introduce background and the importance of antenna optimization. Then the development and trend of antenna optimization based on AI are discussed. In the second chapter, we emphatically introduced the basic concepts and principles of particle swarm algorithms and improved particle swarm algorithms, as well as the flow of their use, following by a brief description of other optimization algorithms such as GA.

Then in Chaps. 3–9, examples of using optimization algorithms to optimize antenna are explained in detailed. All the results mentioned in this book are published journal, articles and patents by our groups. In the third chapter, a multi-frequency antenna is successfully proposed by the steps combining rough designs and precise designs. The condition that the new grids can be only placed near the old one is set to avoid the discontinuity. A-CLPSO is introduced to avoid premature convergence. In the 4th chapter, carved patch antenna part and carved middle ground based on Particle Swarm Optimization (PSO) method is proposed, which can work well on the metal ground at Chinese RFID frequency band—840 to 845 MHz. In the 5th chapters, two parasitic patch are established by grid and optimized by PSO. Placed under the spiral antenna, the antenna requires more wider bandwidth, stable radiation pattern and better circular performance. In 6th chapter, a binary-code mm-wave antenna with broadband, dual-polarization and wide beamwidth is proposed. The driven patch and the parasitic patch are formed by series of small rectangular units optimized by binary codes with multi-objects. The designed mm wave antenna element can work from 21.8 to 33.2 GHz. In 7th chapter, In this chapter, we first use the genetic algorithm to optimize the parameters of the UWB antenna. In 8th chapter, discrete hexagon grids in parasitic layer are presented, which guarantee the line-to-line connection between adjacent elements. After optimizing the hexagon grids, a low-profile antenna with wide CP bandwidth and stable unidirectional pattern is obtained. In the 9th chapter, the configuration of the particle swarm optimization algorithm is presented. And a continuous and smooth structure generating method is described. In the end, the optimized result and experiment data are analyzed.

References

1. D.S. Linden, E.E. Altshuler, Automating wire antenna design using genetic algorithms. Microwave J. **39**(3), 74–86 (1996)
2. J. Holland, in *Adaptation of Natural and Artificial Systems* (1994)
3. M. Ding, R. Jin, J. Geng, G. Yang, Z. Fang, W. He, Design of UWB antennas using GA-FDTD approach, in *2007 IEEE Antennas and Propagation Society International Symposium*, Honolulu, HI, pp. 753–753 (2007). https://doi.org/10.1109/APS.2007.4395603
4. M. Ding, R. Jin, J. Geng, Optimal design of UWB antennas using a mixed model of 2-D GA and FDTD. Microw. Opt. Technol. Lett. **49**(12), 3177–3180 (2007)
5. M. Ding, R. Jin, J. Geng, Auto-design of band-notched UWB antennas using mixed model of 2D GA and FDTD. Electron. Lett. **44**(4), 257–258 (2008)

6. J. Kennedy, R.C. Eberhart, Particle swarm optimization, *in IEEE International Conference on Neural Networks*, vol. 4, pp. 1942–1948 (1995)
7. G. Lu, Z. Cao, Radiation pattern synthesis with improved high dimension PSO, in *2017 Progress in Electromagnetics Research Symposium—Fall (PIERS—FALL)*, Singapore, pp. 2160–2165 (2017)
8. X. Zhao, Y. Jin, H. Ji, J. Geng, X. Liang, R. Jin, An improved mixed-integer multi-objective particle swarm optimization and its application in antenna array design, in *2013 5th IEEE International Symposium on Microwave, Antenna, Propagation and EMC Technologies for Wireless Communications*, Chengdu, pp. 412–415 (2013)
9. H. Wu, J. Geng, R. Jin, J. Qiu, W. Liu, J. Chen, S. Liu, An improved comprehensive learning particle swarm optimization and its application to the semiautomatic design of antennas. IEEE Trans. Antennas Propag. **57**(10), 3018–3028 (2009)
10. J. Geng, X. Liu, R. Jin, X. Liang, Directional wideband antenna with low profile on PEC, in *2012 IEEE International Workshop on Antenna Technology (iWAT)*, Tucson, AZ, 2012, pp. 76–79
11. X. Tao, J. Geng, R. Jin, X. Liang, H. Wu, RFID tag antenna for use on metal, in *2010 International Workshop on Antenna Technology (iWAT)*, Lisbon, pp. 1–4 (2010)
12. L. Lizzi, F. Viani, R. Azaro, A. Massa, Optimization of a spline-shaped UWB antenna by PSO. IEEE Antennas Wirel. Propag. Lett. **6**, 182–185 (2007)
13. H.P. Schwefel, T. Back, Evolution stratcgies I: Tariants and their computational implementation, in *Proceedings of the Genetic Algorithms in Engineering and Computer Science* (Wiley, New York, 1995), pp. 111–126
14. D.B. Fogel, Applying evolutionary programming to selected traveling salesman problem. Cybern. Syst. **24**(1), 27–36 (1993)
15. J.H. Holland, *Adaptation in Natural and Artificial System* (University of Michigan Press, Ann Arbor, 1975)
16. A. Colorini, M. Dorigo, V. Maniczzo, Distributed optimization by ant colonics, in *Proceeding of the First European Conference on Artificial Life* (Elsevier Publishing, Paris, France, 1991), pp. 134–142
17. J. Kennedy, R.C. Eberhart, Particle swarm optimization, in *Proceedings of IEEE International conference on Neural Network* (IEEE, Perth, Australia, 1995), pp. 1942–1948
18. X. Li, Z. Shao, J. Qian, An optimization model based on animal autonomy: fish school algorithm. Syst. Eng. Theory Pract. **11**, 32–38 (2002)
19. S. Kirkpatrick, C.D. Gelatt Jr, M.P. Vecchi, Optimization by simulated annealing. Science **220**, 671–680 (1983)
20. F. Glover, Tabu search-part I. ORSA J. Comput. **1**(3), 190–206 (1989)
21. F. Glover, Tabu search-part II. ORSA J. Comput. **2**(1), 4–32 (1990)
22. J. Zhao, An intelligent optimization method of antenna feeder based on minimization of drive test (MDT) data. Telecommun. Eng. Technol. Standard. **12**, 47–49 (2017)
23. L. Li, Y. Zhang, B. Hu, et al., Intelligent optimization of antenna feeder based on MEC and K-means clustering system and method design of telecommunications. Telecommun. Eng. Technol. Standard. **8** (2019)
24. Q. Zhong, X. Wu, Y. Luo, in *Research based on MDT intelligent analysis of LTE wireless interference Research and Application* [J/OL]. [2019–07–30].http://kns.cnki.net/kcms/detail/11.2103.TN.20190801.1241.004.html
25. Q. Wen, P. Huang, B. Wang, ACP network optimization system based on MDTMR data. China New Commun. **20**(20), 50–51 (2018)
26. Z. Altman, R. Mittra, A. Boag, New designs of ultra wide-band communication antennas using agenetic algorithm. IEEE Trans. Antennas Propag. **45**(10), 1494–1501 (1997)

27. S. Zhang, Application of neural network in antenna design. Modern Electron. Technol. **33**(15), 71–73, 76 (2010)
28. S. Mishra, R.N. Yadav, R.P. Singh, Directivity estimations for short dipole antenna arrays using radial basis function neural networks. IEEE Antennas Wirel. Propag. Lett. **14**, 1219–1222 (2015)
29. X. Liye, *Research on Intelligent Design of Electromagnetic Field Based on Machine Learning* (University of Electronic Science and Technology of China, Chengdu, 2019)

Chapter 2
Binary Coding and Optimization Method

Optimization technology is a powerful tool for modern science and technology. This chapter mainly presents the knowledge about particle swarm algorithms and their improved algorithms.

2.1 Binary Coding and Construct the Antenna

2.1.1 Binary Coding Method

In the traditional antenna design, the structure of the antenna comes from the existing prototypes in the experience, and the optimization of the antenna can only be optimized in the size of the structural parameters of the antenna, such as length, width, spacing and so on. The optimization of these parameters is very helpful for the fine-tuning optimization of the structure and the corresponding antenna performance. However, it is difficult to find a suitable antenna shape in a large range by using this size optimization method. On the other hand, if the maximum boundary area of the antenna can be discretized into tiny metal elements, when the maximum size of the elements is less than a certain infinity, the continuous arrangement of these elements can be approximately considered to be almost the same as the original continuous shape. In this case, we can assign a value of "0" or "1" according to whether there is a metal unit at the location of each unit, where 0 means that the location is a air unit; 1 indicates that there is a metal unit in this position. Then we realize binary code to the area within the antenna size range. It is a binary code sequence to write the code values of all units into a string.

Obviously, different binary coding sequences correspond to different antenna shapes. In other words, such a coding sequence represents a binary digital coding antenna. Figure 2.1a is just the coding sequence of a binary digital coding antenna, and Fig. 2.1b is the antenna shape corresponding to the coding sequence.

J. Geng and R. Jin, *Antenna Optimization and Design Based on Binary Coding*, Modern Antenna, https://doi.org/10.1007/978-981-16-7965-0_2

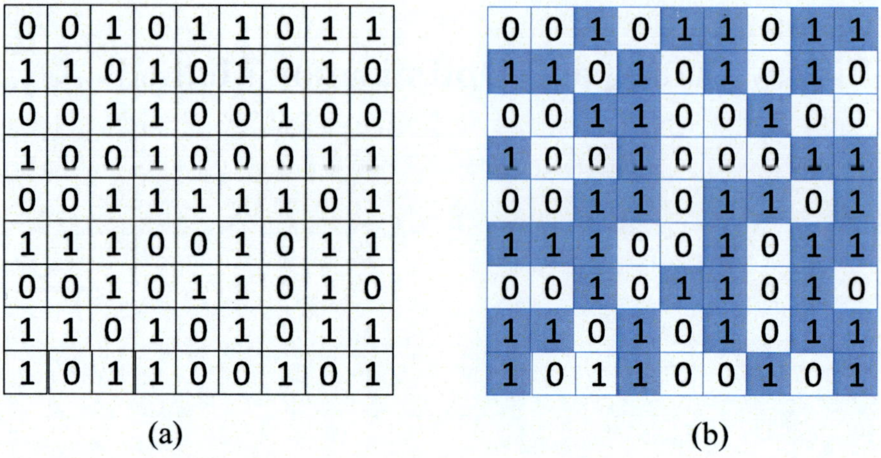

Fig. 2.1 Binary coding to shape: **a** binary coding; **b** corresponding 2D shape

In practice, binary coded shape can represent not only the two-dimensional shape, but also the one-dimensional wire antenna shape and the three-dimensional shape. In this way, in the specific optimization process, for a spatial location, we can take its shape value as a sample, and the value of this sample is "0" or "1".

2.1.2 Construct the Binary Coding Shape from Zero

For the 2D binary coding antenna, although we can map the binary sequence to a specific shape, how to realize it in the specific 3D software modeling?

Firstly, assume that there is no metal in the problem area, and the available area is divided into $M \times M$ micro air units with the thickness of the metal film.

Secondly, a cube with the same size as the unit is constructed directly above the position of each square unit (height *hd*);

Thirdly, the shape value of the position (*i*, *j*) is initialized as

$$w_{\mathrm{i,j}} = random(0.9 \sim 1.1) \tag{2.1}$$

Fourthly, digitize the shape at the position (*i*, *j*), and set

$$z_{\mathrm{i,j}} = int(w_{\mathrm{i,j}}) = \begin{cases} 0 \\ 1 \end{cases} \tag{2.2}$$

Obviously, $z_{\mathrm{i,j}}$ is 0, or 1.

In the fifth step, the cube cell above the position (*i*, *j*) is moved vertically by Boolean operation. The direction of moving is downward and the distance of moving is $z_{i,j} \cdot h_d$. The cube cells above each position are moved vertically by Boolean operation successively.

The sixth step is to multiply the moved cubes in the whole area with the original $M \times M$ micro air units. The result of the product is just the shape of the antenna.

The whole processes to construct the binary coding shape from zero is given in Fig. 2.2.

The above process is starting from 0, which means that the initial material of this area is free space. Then, according to the above steps, these meshes are generated one by one, including the material. The final generated arrangement is the shape of the antenna.

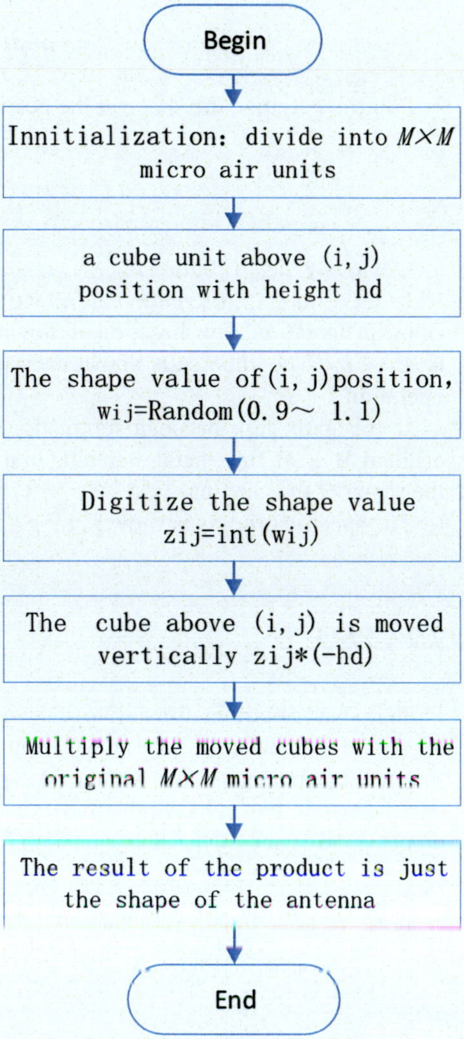

Fig. 2.2 The process to construct the binary coding shape from zero

2.1.3 *Construct the Binary Coding Shape from the Whole Metal*

Besides the above process from 0 to construct the binary coding shape, we can start with a whole piece of metal material, and then subtract it one by one, and the rest is the shape of the antenna.

First of all, we assume that the available area is initially covered with metal, and divide it into $M \times M$ metal units.

In the second step, a cube with the same square as the unit is constructed directly above the position of each square unit (height h_d);

The third step is to initialize the shape of position (i, j) is initialized as

$$\mathrm{aw}_{\mathrm{i,j}} = random(0.9 \sim 1.1) \tag{2.3}$$

Fourthly, digitize the shape at the position (i, j), and set

$$az_{\mathrm{i,j}} = int(aw_{\mathrm{i,j}}) = \begin{cases} 0 \\ 1 \end{cases} \tag{2.4}$$

Obviously, $az_{\mathrm{i,j}}$ is 0, or 1.

In the fifth step, the cube cell above the position (i, j) is moved vertically by Boolean operation. The direction of moving is downward and the distance of moving is $az_{\mathrm{i,j}} \cdot h_d$. The cube cells above each position are moved vertically by Boolean operation.

In the sixth step, Boolean subtraction is performed on the moved cube and the original $M \times M$ tiny metal elements in the whole area. The result of subtraction is the shape of the antenna.

The specific process is shown in Fig. 2.3.

2.2 PSO

Particle swarm optimization is one of the most popular meta heuristic algorithms. The algorithm simulates the flight navigation and foraging behavior of birds. Although the mathematical model is simple, the particle update method is simple and the parameters are few, it has high efficiency in solving complex optimization problems and is widely used in varCious research fields.

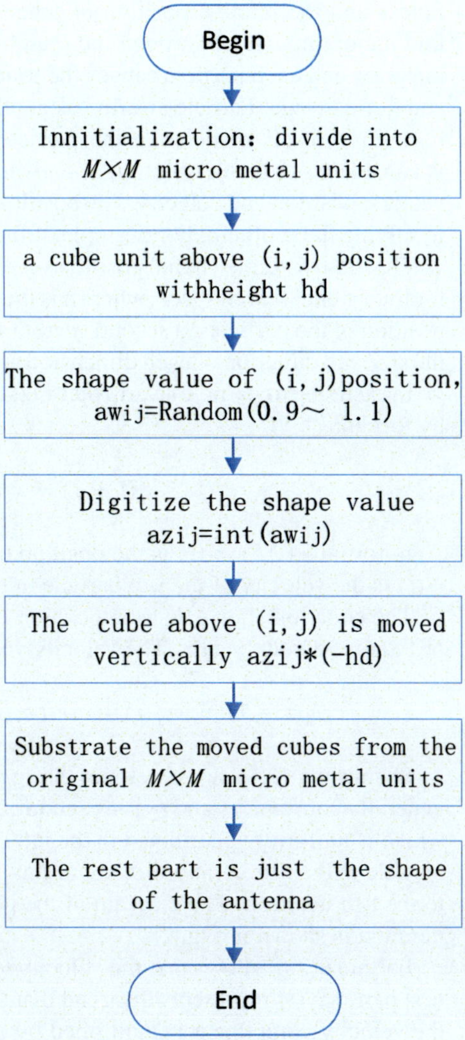

Fig. 2.3 The process to construct the binary coding shape from whole metal

2.2.1 The Basic Principle of Particle Swarm Optimization

In particle swarm optimization (PSO), particles are initialized and then updated to find the optimal solution. Particle swarm optimization is a general optimization algorithm, which uses the social communication knowledge displayed in the animal group to imitate some social behaviors of birds or fish. Similar to genetic algorithm, particle swarm algorithm is also considered as an algorithm that depends on population. The solution provided by the algorithm is regarded as a particle rather than an individual in the genetic algorithm, and the particle swarm in a specific iteration is called a swarm. In PSO, the position of each particle is constantly updated according to the fitness and the relative position of other particles in the particle swarm. Particles find the optimal value through the search space at a dynamic speed. Particles have memory, and each particle carries the trajectory of its previous best position (*p*best) and fitness value. Particle swarm has another value called gbest, which is the global best value of all particles in the particle swarm. In this algorithm, each particle is considered to be the solution of a given optimization problem. It consists of two vectors: position and velocity. The position vector includes the value of each variable in solving the problem. For example, if the problem has two parameters, the particle will have a two-dimensional position vector. Each particle will be able to move in the n-dimensional search space, where n is the number of variables. In order to update the position of the particles, a second vector (velocity) is considered. This vector defines the size and direction of each dimension and each particle update step independently.

In each step of particle swarm optimization, the position of each particle is updated by formula (2.5).

$$x_i(t+1) = x_i(t) + v_i(t+1) \tag{2.5}$$

In formula (2.5), $x_i(t)$ is the position of the i-*th* particle in the t-*th* iteration, and $v_i(t)$ is the velocity of the i-th particle in the t-th iteration.

The equation shows that the position update is simple and the main component of PSO is velocity vector. The velocity vector is defined as follows:

$$v_i(t+1) = wv_i(t) + c_1r_1(p_i(t) - x_i(t)) + c_2r_2(p_g - x_i(t)) \tag{2.6}$$

In formula (2.6), w represents the inertia weight; c_1 and c_2 is learning factor, generally constant; r_1 and r_2 is a pseudo-random number with a value range of [0,1], which is uniformly distributed in the interval; $p_i(t)$ is the best solution from the i-*th* particle to the t-th iteration. p_g represents the best solution found by all particles up to the t-*th* iteration. The diagram of the position update of particle position in each iteration is shown in Fig. 2.4.

Formula (2.6) shows that the velocity vector consists of three components. In the first part, $wv_i(t)$ represents the trend that the particle keeps the present velocity, and the velocity component is multiplied by an inertia weight parameter w. The larger the value of w, the higher the trend of maintaining the previous speed. In the second part, $c_1r_1(p_i(t) - x_i(t))$ represents the best solution obtained by each particle so far to imitate the individual optimization ability of each particle. If the i-*th* particle finds a better solution, the vector $p_i(t)$ is updated every iteration. The influence of learning factor c_1 on the final value of speed can be increased or decreased by changing its range. This parameter is multiplied by a random number in [0,1] to provide random behavior, because particle swarm optimization is a random optimization algorithm. Generally speaking, the second part maintains the trend of the best solution found by individual particles at present, called "personal best solution". In the third part, $c_2r_2(p_g - x_i(t))$ represents the population intelligence that imitates bird swarm, in which all the best solutions obtained by particles are stored in p_g and used in this

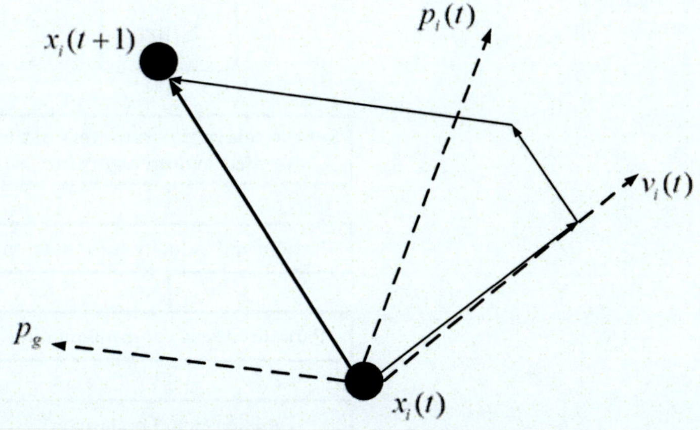

Fig. 2.4 Update position of particle position in each iteration

part. This means that all particles will be attracted to the best solution found by the group. The impact of this part can also be adjusted using c_2.

2.2.2 *The Basic Steps of Particle Swarm Optimization*

Particle swarm optimization algorithm uses the above simple concepts to find the global optimal solution of a given optimization problem. Firstly, the relevant parameters of PSO are set, including population size, maximum iterations, inertia weight w and learning factors c_1, c_2.It starts with a random solution, initializing the positions and velocities of all particles in the particle swarm. Then perform the following steps iteratively until the end condition is met:

Step 1: calculate the fitness value of all particles.

Step 2: update w, c_1, c_2.If c_1 and c_2 are constants, only w is updated.

Step 3: update the individual optimal value and global optimal value.

Step 4: use formula (2.6) to calculate the velocity vector of each particle.

Step 5: use formula (2.5) to calculate the next position of each particle.

Step 6: Judge whether the number of iterations of the particle swarm algorithm has reached the maximum number of iterations or convergence condition, if it is satisfied, then end iterative optimization and proceed to step 7. Otherwise, return to step 1.

Step 7: output the global optimal value.

The overall flow of the particle swarm algorithm is shown in Fig. 2.5.

Fig. 2.5 Flow chart of particle swarm algorithm

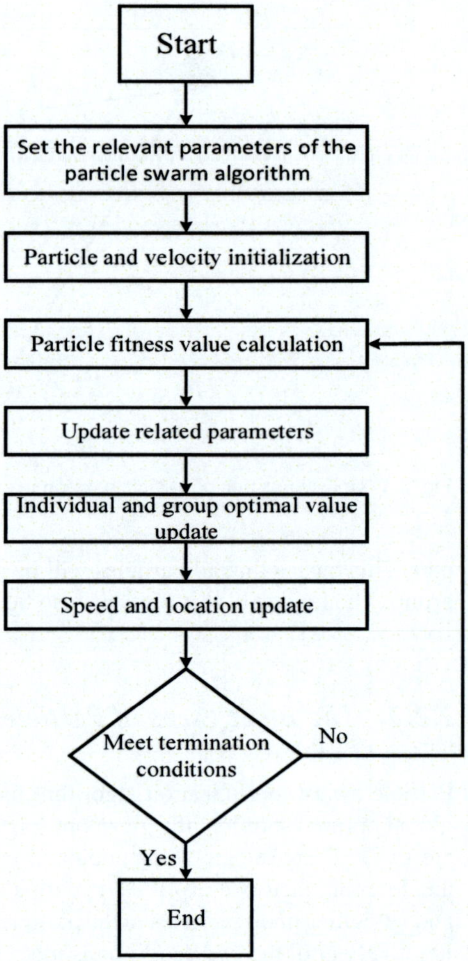

2.3 Improved PSO

Particle swarm optimization algorithm imitates the behavior of birds looking for food, and achieves global optimization by changing the speed of each particle. In the basic particle swarm optimization algorithm, each particle moves towards its previous best position (pbest) and global best position (gbest) [1], but this simple learning behavior can easily to lead to premature. In order to overcome precocity, many improved PSO algorithms have been proposed [2–10]. In these algorithms, CLPSO can deal with most problems more effectively, especially multi extremum problems. After many experiments, we find that the global search ability of CLPSO can be significantly improved if the learning strategies are more diversified. Therefore, based on a more effective learning strategy, A-CLPSO is proposed.

On the other hand, with the rapid development of wireless communications, 3G, WiMax, and WLAN have been increasingly used in life, but the miniaturization and multi-frequency antennas have become a bottleneck restricting the development. Using the effective combination of semi-automatic design method and A-CLPSO, a small multi-frequency antenna is designed whose frequency band covers TD-SCDMA, IEEE 802.11b/g, WiMax and IEEE 802.11a. In the semi-automatic design method, the first step is the rough design of the antenna prototype. Unlike the general design method, each grid is set individually. In the rough design, the grid setting is in a certain order. At the same time, a new objective function is also proposed for the design of multi-frequency antennas. The second step is precise design. The prototype generated in the rough design becomes a more reasonable structure through simple changes. Then the influence of each part on the reflection coefficient is analyzed, and these important parts are optimized. The completion of the two steps depends on the guidance of A-CLPSO. Finally, the designed antenna successfully achieves all the expected goals.

2.3.1 General Particle Swarm Optimization

In general particle swarm optimization, the particles move to pbest and gbest, Learn from the following formulas (2.7) and (2.8).

$$V_i^d \leftarrow w \times V_i^d + c_1 \times rand1_i^d \times \left(pbest_i^d - X_i^d\right) + c_2 \times rand2_i^d \times \left(gbest^d - X_i^d\right) \tag{2.7}$$

$$particle_i^d \leftarrow particle_i^d + V_i^d \tag{2.8}$$

In formula (2.8), $\boldsymbol{particle_i} = (particle_i{}^1, particle_i{}^2, \ldots, particle_i{}^D)$ represents the position of the i-th particle; $\boldsymbol{V_i} = (V_i{}^1, V_i{}^2, \ldots, V_i{}^D)$ represents the velocity of the i-th particle. In formula (2.7), $\boldsymbol{pbest_i} = (pbest_i{}^1, pbest_i{}^2, \ldots, pbest_i{}^D)$ represents the best position reached by the i-th particle; $\boldsymbol{gbest_i} = (gbest_i{}^1, gbest_i{}^2, \ldots, gbest_i{}^D)$ represents the position of the best pbest, that is, the best position reached globally. D represents the dimensionality of the problem. Since all particles only learn from a single pbest and gbest, this kind of algorithm lacking a comprehensive learning strategy can easily lead to premature maturity.

2.3.2 Comprehensive Learning Particle Swarm Algorithm (CLPSO)

In order to avoid premature maturity, CLPSO has adopted a new learning strategy, requiring particles to not only learn their own pbest, but also have a certain probability to learn the pbest of other particles. This probability is called Pc. Therefore, the speed update formula of CLPSO is as follows:

$$V_i^d \leftarrow w \times V_i^d + c \times rand_i^d \times \left(pbest_{fi(d)}^d - X_i^d\right) \quad (2.9)$$

where $f_i = (f_i(1), f_i(2), \ldots, f_i(D))$ represents the particle number that the i-th particle should learn. Different particles are initialized with different Pc values and remain unchanged throughout the search process.

2.3.3 *Disadvantages of CLPSO*

Many test functions show that CLPSO has stronger search capabilities than other improved particle swarm algorithms, but for some functions, such as Griewanks and rotated Griewanks functions, CLPSO's learning strategy is still too rigid to solve these functions well [9]. Experiments show that CLPSO has 3 potential disadvantages:

(1) It is not always reasonable to always choose the winner in the competition as the learning object. Sometimes choosing the loser to learn can help jump out of the local optimum.
(2) The Pc of each particle is fixed at the beginning, and the Pc of some particles may be inappropriate. For example, if a poor particle has a small Pc, it is difficult for it to learn from other particles to reach a good position; Conversely, a good particle with a large Pc is likely to destroy their own convergence properties due to frequent learning Inferior particles.
(3) Once all particles fall into local convergence, CLPSO cannot provide an effective strategy to make them jump out.

2.3.4 *Adaptive-Comprehensive Learning Particle Swarm Algorithm (A-CLPSO) [11]*

In order to solve the three shortcomings of CLPSO, the following three improvement strategies are proposed to solve the above three shortcomings. The following experimental data show that these three improvement strategies are indispensable for improving the performance of the algorithm.

2.3.4.1 Choose the Time Period of the Loser

In order to solve shortcoming 1, in A-CLPSO, the particles will choose the loser to learn in a certain period of time, which is when $times \in [g\_start, g\_end]$ (1 <

Table 2.1 Results for different setting time periods

Type	Function	
	Griewanks	Rotated Griewanks
①	4.93e−004 ± 2.15e−003	2.16e−007 ± 9.44e−007
②	3.71e−004 ± 1.61e−003	6.62e−005 ± 9.62e−005
③	4.24e−006 ± 1.85e−005	2.04e−007 ± 8.65e−007
④	3.70e−004 ± 1.61e−003	8.32e−006 ± 3.61e−005
⑤	**2.75e−013 ± 8.20e−013**	**1.48e−014 ± 6.43e−014**
⑥	3.25e−004 ± 1.33e−003	1.27e−010 ± 5.57e−010

g_start < g_end < nc). It is found that different settings of 'g_start' and 'g_end' have a greater impact on the performance of the algorithm.

In J. J. Liang's article [9], CLPSO cannot solve the Griewanks and rotated Griewanks functions. Therefore, these two functions are now used to evaluate the quality of different settings of g_start and g_end. The following 6 settings are mainly set:

① Don't set the time period for choosing losers.
② g_start = 1, g_end = nc.
③ g_start = 1, g_end = $\frac{nc}{2}$.
④ g_start = 1, g_end = $\frac{nc}{5}$.
⑤ Set two time periods: $g\_start_1 = 1$, $g\_end_1 = \frac{1}{10}nc$, $g\_start_2 = \frac{6}{10}nc$, $g\_end_2 = \frac{7}{10}nc$.
⑥ Set three time periods: $g\_start_1 = 1$, $g\_end_1 = \frac{1}{10}nc$, $g\_start_2 = \frac{4}{10}nc$, $g\_end_2 = \frac{5}{10}nc$, $g\_start_3 = \frac{7}{10}nc$, $g\_start_3 = \frac{8}{10}nc$.

The dimension of the function is 10, the number of particles (particle_num) is set to 30 and NC is set to 1000. For each function operation, repeat 30 times. The average and variance are listed in Table 2.1. The results show that the effect is the best when the time period is set using ⑤.

2.3.4.2 Dynamic Update of Pc

In order to solve shortcoming 2, A-CLPSO proposes a dynamic Pc value, which allows good particles to focus more on their own search (their Pc value is set to be smaller), and allows poor particles to learn more other particles (Their Pc value is set to be larger).

The dynamic value of Pc is generated according to the following method: after each iteration, all the particles are arranged from good to bad according to their pbest value, and then Pc is set according to the sequence number in the arrangement. The Pc of the i-th particle is set:

Table 2.2 Results of static Pc and dynamic Pc

Type	Function	
	Griewanks	Rotated Griewanks
Static Pc	4.23e−004 ± 1.61e−003	5.71e−005 ± 2.38e−004
Dynamic Pc	**2.75e−013 ± 8.20e−013**	**1.48e−014 ± 6.43e−014**

$$Pc_i = 0.05 + 0.35 \times \left(\frac{f(i)}{particle\_num}\right)^2 \tag{2.10}$$

In formula (2.10), f(i) is the number of the i-th particle in the arrangement. Empirically speaking, particles should pursue faster convergence in the early stage of the search, and should pursue more interactions in the later stage. Therefore, after each iteration, if a better gbest is not found, each Pc will be increased by a small amount shown in formula (2.11).

$$Pc_i = Pc_i + 0.0001 \times i^{0.2} \tag{2.11}$$

The Griewanks and rotated Griewanks functions continue to be used to show the difference between using static Pc and dynamic Pc. The results in Table 2.2 show that the use of dynamic Pc can significantly improve the algorithm performance.

2.3.4.3 Velocity Disturbance

In order to solve the shortcoming 3, considering the speed update formula (2.9) of CLPSO, if all particles fall into the local optimum, this update formula cannot provide the motivation to make them jump out. Therefore, a velocity disturbance that is inversely proportional to the degree of particle aggregation is proposed in this section. First, define mbest as the center of all pbest:

$$\boldsymbol{mbest} = (mbest^1, mbest^2, \ldots, mbest^D) \tag{2.12}$$

$$mbest^d = \frac{1}{particle\_num} \sum_{i=1}^{particle\_num} pbest_i^d \tag{2.13}$$

In this way, the distance between the i-th particle and mbest on the d-th dimension can be expressed as $\left|mbest^d - particle_i^d\right|$, Unlike pbest and gbest, which always attract particles, mbest always pushes particles away from itself. The closer the particles are, the farther they will be rebounded. This speed disturbance is expressed as the following formula:

$$V_i^d \leftarrow V_i^d - rand_i^d \times \frac{m}{mbest^d - particle_i^d} \tag{2.14}$$

Table 2.3 Results for different values of m and dist

dist	*m*	Griewanks	Rotated Griewanks
$0.02 \times \sqrt{\frac{times}{nc}}$	0.0005	5.20e−012 ± 2.12e−011	7.23e−006 ± 4.72e−005
	0.005	6.16e−004 ± 2.68e−003	4.93e−004 ± 2.15e−003
	0.05	3.70e−004 ± 1.61e−003	1.23e−003 ± 2.97e−003
$0.05 \times \sqrt{\frac{times}{nc}}$	0.0005	7.40e−004 ± 2.22e−003	8.63e−004 ± 2.62e−003
	0.005	8.57e−008 ± 3.74e−007	5.19e−004 ± 2.15e−003
	0.05	**2.75e−013 ± 8.20e−013**	**1.48e−014 ± 6.43e−014**
$0.1 \times \sqrt{\frac{times}{nc}}$	0.0005	2.20e−012 ± 8.90e−012	9.89e−004 ± 3.06e−003
	0.005	4.93e−004 ± 2.15e−003	6.26e−014 ± 2.73e−013
	0.05	3.70e−004 ± 1.61e−003	2.09e−004 ± 9.12e−004

The value of m determines the severity of the disturbance. The speed disturbance occurs with a certain probability, which is called 'dist'. A lot of experiments on different values of m and dist has done in this section. The results are listed in Table 2.3, which shows that parameters when m = 0.05 and dist = $0.05 \times \sqrt{\frac{times}{nc}}$,are more appropriate.

2.3.4.4 Principal Block Diagram of A-CLPSO

Three improvements of A-CLPSO have been discussed. As shown in Fig. 2.6, the block diagram of the algorithm is as follows:

particle_num: Number of particles.
nc: Maximum iterated algebra.
times: Iterations, from 1 to NC.
i: Particle number, from 1 to particle_num.
d: dimension.
$\text{particle}_i{}^d$: The position of the i-th particle in the d-th dimension

$$w = w_{\max} - \frac{times}{nc}(w_{\max} - w_{\min}) \tag{2.15}$$

Here, $w_{\max} = 0.6$, $w_{\min} = 0$,$c = \begin{cases} 0.5 + rand & fi(d) \neq i \\ 0.9 + 0.2 \times rand & fi(d) = i \end{cases}$.

2.3.4.5 Test Results

(1) Test function

12 test functions are used to compare A-CLPSO with other improved PSO in this section. The 12 functions are as follows:

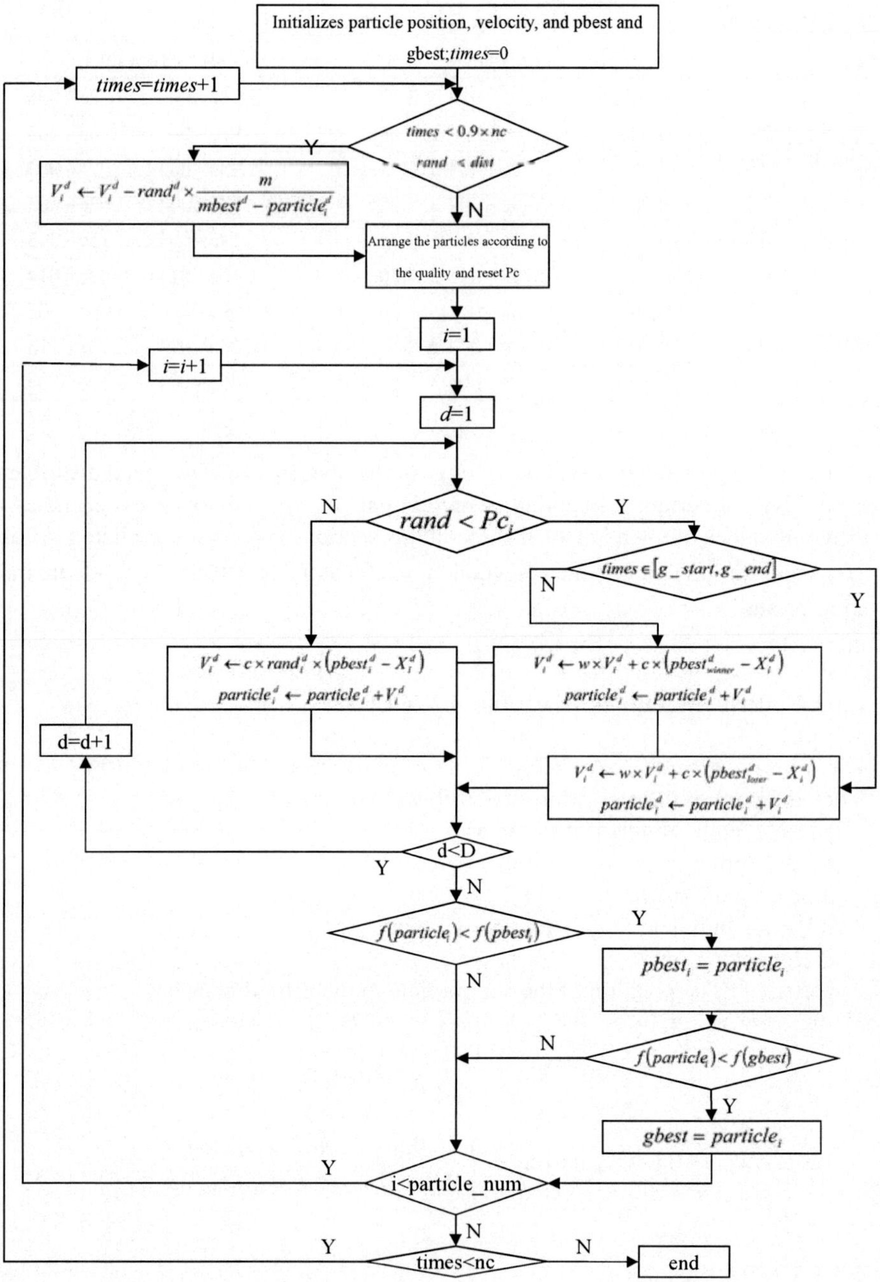

Fig. 2.6 Principal block diagram of A-CLPSO

(1) Sphere function

$$f_1(x) = \sum_{i=1}^{D} x_i^2 \tag{2.16}$$

(2) Rosenbrock's function

$$f_2(x) = \sum_{i=1}^{D-1} \left(100\left(x_i^2 - x_{i+1}\right)^2 + (x_i - 1)^2\right) \tag{2.17}$$

(3) Ackley's function

$$f_3(x) = -20\exp\left(-0.2\sqrt{\frac{1}{D}\sum_{i=1}^{D} x_i^2}\right) - \exp\left(\frac{1}{D}\sum_{i=1}^{D}\cos(2\pi x_i)\right) + 20 + e \tag{2.18}$$

(4) Griewanks's function

$$f_4(x) = \sum_{i=1}^{D} \frac{x_i^2}{4000} - \prod_{i=1}^{D} \cos\left(\frac{x_i}{\sqrt{i}}\right) + 1 \tag{2.19}$$

(5) Weierstrass function

$$f_5(x) = \sum_{i=1}^{D}\left(\sum_{k=0}^{k\max}\left[a^k \cos\left(2\pi b^k (x_i + 0.5)\right)\right]\right) - D\sum_{k=0}^{k\max}\left[a^k \cos\left(2\pi b^k \cdot 0.5\right)\right] \tag{2.20}$$

$$a = 0.5\ b = 3\ k\max = 20$$

(6) Rastrigin's function

$$f_6(x) = \sum_{i=1}^{D}\left(x_i^2 - 10\cos(2\pi x_i) + 10\right) \tag{2.21}$$

(7) Discontinuous Rastrigin's function

$$f_7(x) = \sum_{i=1}^{D}\left(y_i^2 - 10\cos(2\pi y_i) + 10\right)$$

$$y_i = \begin{cases} x_i & |x_i| < 0.5 \\ 0.5 \times round(2x_i) & |x_i| \geq 0.5 \end{cases} \tag{2.22}$$

(8) Schwefel's function

$$f_8(x) = 418.9829 \times D - \sum_{i=1}^{D} x_i \sin\left(\sqrt{|x_i|}\right) \tag{2.23}$$

(9) Rotated Ackley's function

$$f_9(x) = -20 \exp\left(-0.2\sqrt{\frac{1}{D}\sum_{i=1}^{D} y_i^2}\right) - \exp\left(\frac{1}{D}\sum_{i=1}^{D}\cos(2\pi y_i)\right) + 20 + e \quad y = M * x \tag{2.24}$$

(10) Rotated Griewanks's function

$$f_{10}(x) = \sum_{i=1}^{D} \frac{y_i^2}{4000} - \prod_{i=1}^{D} \cos\left(\frac{y_i}{\sqrt{i}}\right) + 1 \quad y = M * x \tag{2.25}$$

(11) Rotated Rastrigin's function

$$f_{12}(x) = \sum_{i=1}^{D} \left(y_i^2 - 10\cos(2\pi y_i) + 10\right) \quad y = M * x \tag{2.26}$$

(12) Rotated Discontinuous Rastrigin's function

$$f_{13}(x) = \sum_{i=1}^{D} \left(z_i^2 - 10\cos(2\pi z_i) + 10\right)$$

$$z_i = \begin{cases} y_i & |y_i| < 0.5 \\ 0.5 \times round(2y_i) & |y_i| \geq 0.5 \end{cases} \quad y = M * x \tag{2.27}$$

M matrix is generated by Salomon's method [12]. The global optimal $\boldsymbol{x}^*$ of all test functions, the corresponding optimal function value, search range and initial offset are listed in Table 2.4. All the test functions are 10 dimensions.

(2) **Comparison between different algorithms**

In this section, A-CLPSO is compared with UPSO, FDR-PSO, FIPS, CPSO-H, CLPSO [9, 13–16] according to the same number of operations. The particle numbers of A-CLPSO and CLPSO are 10 and 30, respectively. The particle numbers of other

Table 2.4 Global optimization of each test function, search range and initial offset

Function	x^*	$f(x^*)$	Search range	Initialization range
f_1	[0, 0,..., 0]	0	$[-100, 100]^D$	$[-100, 50]^D$
f_2	[1, 1,..., 1]	0	$[-2.048, 2.048]^D$	$[-2.048, 2.048]^D$
f_3	[0, 0,..., 0]	0	$[-32.768, 32.768]^D$	$[-32.768, 16]^D$
$f4$	[0, 0,..., 0]	0	$[-600, 600]^D$	$[-600, 200]^D$
f_5	[0, 0,..., 0]	0	$[-0.5, 0.5]^D$	$[-0.5, 0.2]^D$
f_6	[0, 0,..., 0]	0	$[-5.12, 5.12]^D$	$[-5.12, 2]^D$
f_7	[0, 0,..., 0]	0	$[-5.12, 5.12]^D$	$[-5.12, 2]^D$
f_8	[420.96, 420.96,..., 420.96]	0	$[-500, 500]^D$	$[-500, 500]^D$
f_9	[0, 0,..., 0]	0	$[-32.768, 32.768]^D$	$[-32.768, 16]^D$
f_{10}	[0, 0,..., 0]	0	$[-600, 600]^D$	$[-600, 200]^D$
f_{11}	[0, 0,..., 0]	0	$[-5.12, 5.12]^D$	$[-5.12, 2]^D$
f_{12}	[0, 0,..., 0]	0	$[-5.12, 5.12]^D$	$[-5.12, 2]^D$

algorithms are 10. Each algorithm is repeated 30 times for each test function, and the average value and variance are listed in Table 2.5.

From the above results, we can see that CLPSO pays more attention to the improvement of search ability, so it runs better than other algorithms on multiple extremum problems (as can be seen from functions 5, 6, 7 and 8). A-CLPSO inherits this feature of CLPSO. However, according to the no free lunch theory [17], the search ability and convergence rate are often contradictory.

A-CLPSO does not sacrifice the convergence speed too much but makes the learning strategy more effective to improve the search ability, and further improves the search ability of the algorithm (as can be seen from functions 4, 9, 10).

2.4 Genetic Algorithm (GA)

The Genetic algorithm (GA) is a kind of randomized search algorithm proposed by Professor J. Holland of the United States that draws on the natural selection and natural genetic mechanism of the biological world. It is especially suitable for dealing with complex and nonlinear problems that are difficult to solve by traditional search methods, and can be widely used in fields such as combinatorial optimization, machine learning, adaptive control, planning and design, and artificial life. For different problems, many scholars have designed many different coding methods to represent the feasible solutions of the problem, and developed different genetic operators to imitate the genetic characteristics of organisms in different environments. In this way, different encoding methods and different genetic operators constitute different genetic algorithms. However, their common feature is that they all imitate

Table 2.5 Results of each test function

PSOs	Function			
	(1)	(2)	(3)	(4)
UPSO	9.84e−118 ± 3.56e−117	1.40e000 ± 1.88e000	1.33e000 ± 1.48e000	1.04e−001 ± 7.10e−002
FDR-PSO	2.21e−090 ± 9.88e−090	8.67e−001 ± 1.63e000	3.18e−014 ± 6.40e−014	9.24e−002 ± 5.61e−002
FIPS	3.15e−030 ± 4.56e−030	2.78e000 ± 9.88e−001	3.75e−015 ± 2.13e−014	1.31e−001 ± 9.32e−002
CPSO-H	4.98e−045 ± 1.00e−044	1.53e000 ± 1.70e000	1.49e−014 ± 6.97e−015	4.07e−002 ± 2.80e−002
CLPSO (10 particles)	5.15e−029 ± 2.16e−028	2.46e000 ± 1.70e000	4.32e−014 ± 6.97e−015	4.56e−003 ± 4.81e−003
CLPSO (30 particles)	1.21e−023 ± 1.63e−023	5.23e000 ± 6.88e−001	2.73e−008 ± 1.22e−008	6.96e−002 ± 2.59e−002
A-CLPSO (10 particles)	1.18e−023 ± 4.86e−023	5.81e000 ± 2.01e000	1.03e−012 ± 3.84e−012	1.74e−002 ± 1.47e−002
A-CLPSO (30 particles)	7.15e−042 ± 5.78e−041	3.21e000 ± 1.18e000	3.73e−015 ± 7.74e−016	2.75e−013 ± 8.20e−013
	(5)	(6)	(7)	(8)
UPSO	1.14e000 ± 1.17e000	1.17e001 ± 6.11e000	5.85e000 ± 3.15e000	1.08e003 ± 2.63e002
FDR-PSO	3.01e−003 ± 7.20e−003	7.51e000 ± 3.05e000	3.35e000 ± 2.01e000	8.51e002 ± 2.76e002
FIPS	2.02e−003 ± 6.40e−003	2.12e000 ± 1.33e000	4.35e000 ± 2.80e000	7.10e001 ± 1.50e002
CPSO-H	1.07e−015 ± 1.67e−015	0 ± 0	2.00e−001 ± 4.10e−001	2.13e002 ± 1.41e002
CLPSO (10 particles)	0 ± 0	0 ± 0	0 ± 0	0 ± 0

(continued)

Table 2.5 (continued)

PSOs	Function			
CLPSO (30 particles)	0 ± 0	5.19e000 ± 1.86e000	7.16e000 ± 8.58e−001	2.10e002 ± 1.77e002
A-CLPSO (10 particles)	0 ± 0	4.97e−001 ± 5.89e−001	2.30e000 ± 1.23e000	0 ± 0
A-CLPSO (30 particles)	0 ± 0	0 ± 0	0 ± 0	0 ± 0
	(9)	(10)	(11)	(12)
UPSO	1.00e000 ± 9.27e−001	7.76e−002 ± 6.40e−002	1.52e001 ± 5.25e000	1.47e001 ± 6.53e000
FDR-PSO	1.40e−001 ± 4.38e−001	1.44e−001 ± 7.84e−002	9.25e000 ± 2.50e000	1.07e001 ± 3.86e000
FIPS	2.25e−015 ± 1.54e−015	1.70e−001 ± 1.26e−001	1.20e001 ± 6.22e000	8.84e000 ± 3.27e000
CPSO-H	1.36e000 ± 8.85e−001	1.20e−001 ± 8.07e−002	2.67e001 ± 1.06e001	1.70e001 ± 9.05e000
CLPSO (10 particles)	3.56e−005 ± 1.57e−004	4.50e−002 ± 3.08e−002	5.97e000 ± 2.88e000	5.44e000 ± 1.39e000
CLPSO (30 particles)	9.18e−008 ± 2.81e−007	1.40e−001 ± 4.14e−002	1.75e001 ± 2.75e000	1.11e001 ± 2.33e000
A-CLPSO (10 particles)	6.39e−002 ± 2.52e−001	1.39e−002 ± 1.51e−002	5.47e000 ± 2.73e000	6.04e000 ± 2.80e000
A-CLPSO (30 particles)	5.51e−015 ± 3.97e−015	1.48e−014 ± 6.43e−014	1.72e000 ± 8.32e−001	2.19e000 ± 2.24e000

the selection, crossover and mutation of biological mechanisms to complete the adaptive search process for optimal solutions to different problems. Based on this feature, Gofdberg has summarized a unified and most basic genetic algorithm—Simple Genetic Algorithms, referred to as SGA [18]. The basic genetic algorithm only uses the three basic genetic operators of selection, crossover and mutation. Its evolutionary process is simple and easy to understand, which is the rudiment and foundation of some other genetic algorithms. It not only provides a basic framework for various genetic algorithms, but also has certain application value [19].

There are 5 basic components of genetic algorithm [20]:

(1) The genetic representation of the problem solution;
(2) The method of creating the initial population;
(3) Evaluation function for judging the merits of individuals based on their fitness values;
(4) Used to change the individual genetic operators generated during the replication process;
(5) The parameter value of the genetic algorithm.

The genetic algorithm maintains a population P(t) composed of a group of individuals, which representing a potential solution to the problem. (t stands for genetic algebra). Certain individuals undergo random transformations called genetic manipulations, resulting in new individuals. There are two main transformation methods: the mutation method is to change an individual to obtain a new individual; the crossover method is to combine the relevant parts of two individuals to form a new individual. The newly generated individuals C(t) (called offspring) continue to be evaluated for their merits. A new population is formed by selecting superior individuals from the parent population and the offspring population. After several generations, the algorithm converges to an optimal individual, which is likely to represent the optimal or suboptimal solution to the problem. The general structure of genetic algorithm can be described as shown in Fig. 2.7.

2.5 Summary

This chapter mainly introduces the binary coding method and some optimization methods, and it includes the key method of the binary coding and the basic principle and steps of particle swarm optimization, improved PSO, which are the fundamentals and tools behind.

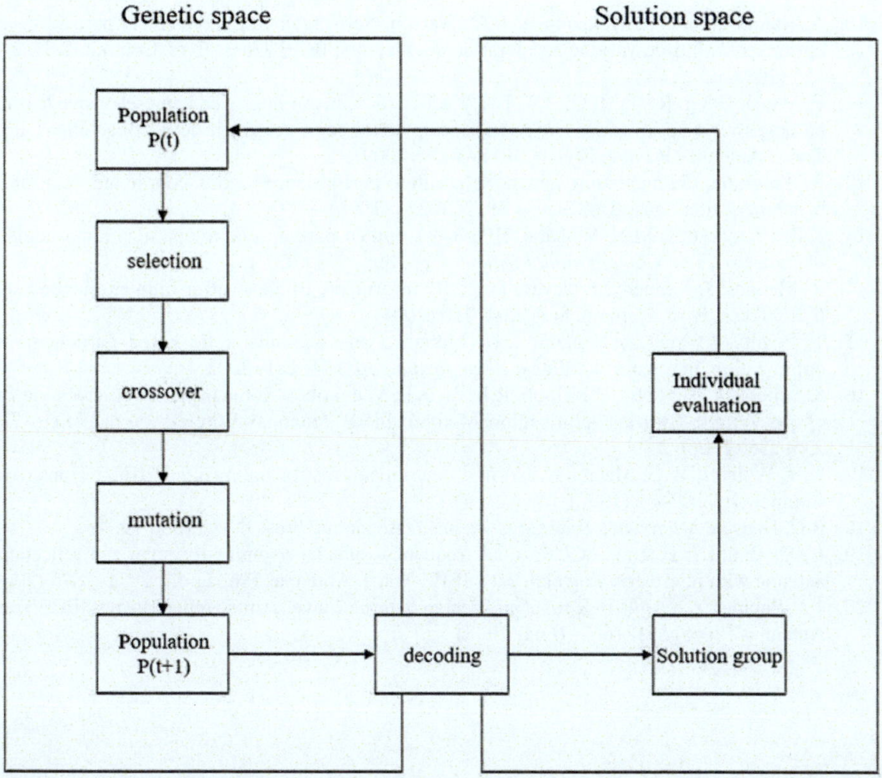

Fig. 2.7 The principle chart of genetic algorithm

References

1. J. Kennedy, R. Eberhart, Particle swarm optimization, in *IEEE International Conference on Neural Networks*, Perth, Australia, vol. 4, pp. 1942–1948 (1995)
2. F. van den Bergh, A.P. Engelbrecht, A cooperative approach to particle swarm optimization. IEEE Trans. Evol. Comput. **8**(3), 225–239 (2004)
3. S. Cui, D.S. Weile, Application of parallel particle swarm optimization scheme to the design of electromagnetic absorbers. IEEE Trans. Antennas Propag. **53**(11), 3616–3624 (2005)
4. S.M. Mikki, A.A. Kishk, Quantum particle swarm optimization for electromagnetics. IEEE Trans. Antennas Propag. **54**(10), 2764–2775 (2006)
5. S. Genovesi, A. Monorchio, R. Mittra, G. Manara, A sub-boundary approach for enhanced particle swarm optimization and its application to the design of artificial magnetic conductors. IEEE Trans. Antennas Propag. **55**(33), 766–770 (2007)
6. T. Huang, A.S. Mohan, A microparticle swarm optimizer for the reconstruction of microwave images. IEEE Trans. Antennas Propag. **55**(3), 568–576 (2007)
7. T. Huang, A.S. Mohan, A hybrid boundary condition for robust particle swarm optimization. IEEE Antennas Wireless Propag. Lett. **4**, 112–117 (2005)
8. S. Selleri, M. Mussetta, P. Pirinoli, R.E. Zich, L. Matekovits, Differentiated meta-PSO methods for array optimization. IEEE Trans. Evol. Comput. **56**(1), 67–75 (2008)
9. J.J. Liang, Comprehensive learning particle swarm optimizer for global optimization of Multimodal Functions. IEEE Trans. Evol. Comput. **10**(3), 281–295 (2006)

10. A. Ratnaweera, S.K. Halgamuge, H.C. Watson, Self-organizing hierarchical particle swarm optimizer with time-varying acceleration coefficients. IEEE Trans. Evol. Comput. **8**(3), 240–255 (2004)
11. H. Wu, J. Geng, R. Jin, J. Qiu, W. Liu, J. Chen, S. Liu, An improved comprehensive learning particle swarm optimization and its application to the semi-automatic design of antennas. IEEE Trans. Antennas Propag. **57**(10), 3018–3028 (2009)
12. R. Salomon, Reevaluating genetic algorithm performance under coordinate rotation of benchmark functions. BioSystems **39**, 263–278 (1996)
13. K.E. Parsopoulos, M.N. Vrahatis, UPSO—A unified particle swarm optimization scheme, in *Lecture Series on Computational Sciences*, pp. 868–873 (2004)
14. R. Mendes, J. Kennedy, J. Neves, The fully informed particle swarm: simpler, maybe better. IEEE Trans. Evol. Comput. **8**(3), 204–210 (2004)
15. T. Peram, K. Veeramachaneni, C.K. Mohan, Fitness-distance-ratio based particle swarm optimization, in *Swarm Intelligence Symposium*, pp. 174–181 (2003)
16. S. Selleri, M. Mussetta, P. Pirinoli, R.E. Zich, L. Matekovits, Some insight over new variations of the Particle Swarm Optimization Method. IEEE Antennas Wireless Propag. Lett. **5**(1), 235–238 (2006)
17. D.H. Wolpert, W.G. Macready, No free lunch theorems for optimization. IEEE Trans. Evol. Comput. **1**(1), 67–82 (1997)
18. R.C. Hansen, *Microwave Scanning Antennas* (Academic Press, New York, 1964)
19. H. Chulmin, J. Huang, AC/ Ka dual frequency dual layer circularly polarized reflectarray antenna with microstrip ring elements. IEEE Trans. Antennas Propag. **52**, 2871–2875 (2004)
20. J.A. Encinar, Z.J. Agustin, Broadband design of three 2 layer printed reflect arrays. IEEE Trans. Antennas Propag. **51**, 1662–1664 (2003)

Chapter 3
Multi Frequency Antenna Design Based on Binary Coding

3.1 Introduction

Generally speaking, the traditional antenna design mainly depends on the design experience of antenna engineers, especially the initial selection of antenna is often closely related to the knowledge, experience and habits of designers [1]. Although a lot of antenna design methods called automatic design have been proposed, the optimization algorithm is mainly used for the size parameters of the prototype structure, such as genetic algorithm (GA), combined with electromagnetic calculation means for repeated calculation and parameter iterative optimization. The final structure often improves the prototype antenna, which is discontinuous [2, 3]. In addition, genetic algorithm is a global optimization algorithm. In the antenna optimization design based on genetic algorithm, multiple objectives need to be considered, and the corresponding fitness function is similar to a multimodal problem, so the optimization result is not necessarily the best. Compared with genetic algorithm, many test functions show that particle swarm optimization (PSO) algorithm may be more effective. This chapter focuses on a semi-automatic antenna design method based on the combination of improved PSO algorithm and new design strategy.

Even for the method itself, the standard particle swarm optimization algorithm has high computational efficiency and fast convergence, but it is easy to lead to premature convergence of multimodal problems [4]. Many researchers have proposed many improved particle swarm optimization algorithms to solve the premature convergence problem [5–13]. CLPSO is an effective method to solve the problem of premature convergence in many improved particle swarm optimization algorithms. However, the optimization of CLPSO algorithm is not very ideal for some functions which fluctuate violently and have many local optimal solutions. On the basis of combing the early PSO algorithm and its improved algorithm, we improved CLPSO algorithm in Chapt.2, which inherits many advantages of CLPSO algorithm, and introduces a more effective learning strategy, which greatly improves the efficiency of the optimization algorithm, we call it A-CLPSO algorithm, The test results show that the A-CLPSO algorithm can optimize multimodal problems more effectively.

J. Geng and R. Jin, *Antenna Optimization and Design Based on Binary Coding*, Modern Antenna, https://doi.org/10.1007/978-981-16-7965-0_3

In order to get rid of too much human experience interference, the antenna optimization design based on binary coding and A-CLPSO algorithm in this chapter includes two steps: coarse model design and fine model design. The first step is the coarse model design, which automatically creates the required antenna prototype. Different from the traditional method of setting grids separately, the binary grids in this chapter are grown one by one in a certain order to avoid discontinuities. The second step is fine model design, which is to smooth the antenna prototype designed in the first step to make its structure more smooth, continuous and reasonable. Then, the key dimensions that affect the performance of the antenna are refined and optimized. These two steps are completed with the help of the proposed A-CLPSO optimization algorithm. Finally, a small multi band monopole antenna covering TD-SCDMA (1.88–2.025 GHz), ieee802.11b/g (2.4–2.484 GHz), WiMAX Application (3.4–3.6 GHz) and IEEE802.11a (5.15–5.85 GHz) is optimized based on the above methods, which also proves the effectiveness of the semi-automatic design method.

3.2 Semi-automatic Design of Antenna with the Aid of A-clpso

In this kind of semi-automatic antenna design process, the process that the designer foresees or gives the antenna prototype based on experience is replaced by the coarse model design, and the modification of the prototype requires little human experience. Furthermore, the accurate model design can ensure that the characteristics of the antenna meet the design requirements.

3.2.1 Grid Setting in Rough Design

In the early optimization algorithm, mesh segmentation is also used, but this simple segmentation is prone to breakpoints or singular points, which can not guarantee the continuity of the structure. Generally, continuity check is carried out after the structure is generated, and it will waste a lot of time to reconstruct the discontinuous parts or singular points. In order to generate continuous structure, we propose a new method of setting grids in a certain order. First of all, given a square area, suppose that this area is divided into 10 parts $\times$ 10 blocks, and 10 dimensional a-clpso algorithm is used for optimization. Correspondingly, a 90 bit binary string is used to map the grid settings (why we only use 90 bit string to map 100 grid settings, which will be discussed later). In order to get this string, the search range in each dimension is set to [0, 512], and the position of the particle in each dimension can be written as a 9-bit binary number (the position is first transformed into an integer from 0 to 511 by an operator "int()". After changing the values of the first dimension to the tenth dimension one by one, a string composed of 90 binary digits is obtained. Then set the grid to "0" or "1" as follows:

Step 1, as shown in Fig. 3.1a, initialize the design area, and divide the design area into n * n uniform grids, assign "1" to the four grids in the center of the area, and set "null" to the other grids. The four grids assigned "1" in the center of the region are named A, B, C and D according to the counter clock, which can also be called "letter" grids. Grid A is defined as the parent grid of grid B and used as the coordinate reference in the next step. Grid B is the parent grid of grid C, grid C is the parent grid of grid D, and grid D is the parent grid of grid A.

Step 2, as shown in Fig. 3.1b, starting from grid A, the direction of grid A pointing to its parent grid is defined as "North". If grid A is not assigned a value (still "Null"), then the "East", "South" and "West" grids of grid A are assigned a numerical values in the binary string one by one. If the grid has been assigned a value of "0" or "1", the corresponding number in the binary string will be retained and the next grid will be assigned from that number. Then, grid A is defined as the parent mesh of those meshes generated by mesh A in this step.

Step 3, as shown in Fig. 3.1c, assign values to all grids around grids B, C and D in the same way as in step 2.

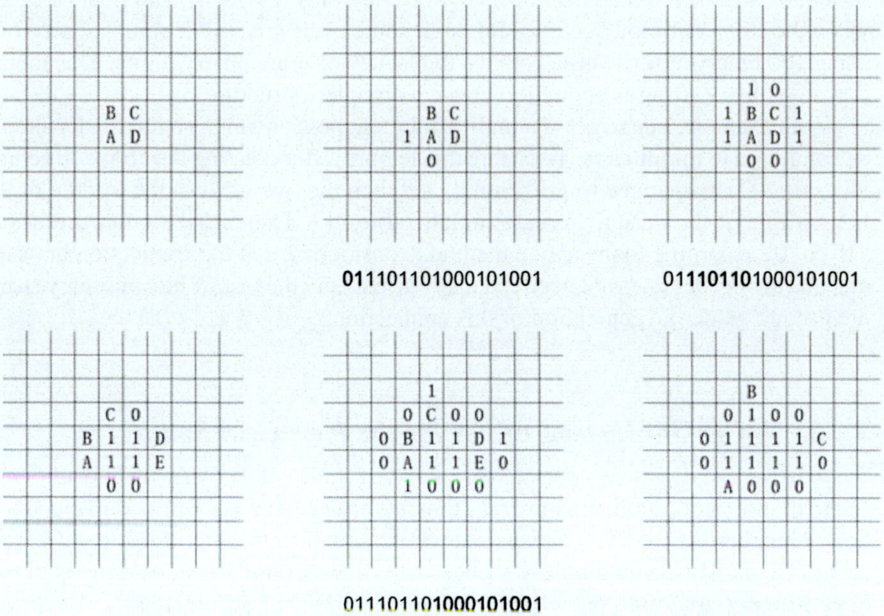

Fig. 3.1 The structure construction process (the digits being used for the grid setting in current Fig are shown in bold): **a** Step 1; **b** Step 2. When setting the third grid around A which is already a grid of "1" (it is named B but it has been set to "1" in the initialization), the third digit is reversed and the next setting starts on it; **c** Step 3; **d** Step 4; **e** Step 2, 3 in the second circulation. **f** Step 4 in the second circulation and then it is ready for the third circulation. Figure reproduced with permission from Ref. [15], © 2009 IEEE

Step 4, as shown in Fig. 3.1d, for the newly generated grid assigned with “1”, name it as a, B, C... According to the order generated in step 2 and step 3, and then cancel the names of all previous “letter” grids. Now, the characteristics of the corresponding design area return to step 1, that is, there are some “letter” grids, and each new “letter” grid has its own parent grid to guide the assignment of other grids. So the next loop goes back to step 2.

During the whole operation, repeat steps 2, 3 and 4 until any of the following conditions occurs:

① All binary strings are read
② New mesh ‘1’ was not created.

Figure 3.1 shows an example, which shows that the position of the particle in the 10 dimensional a-clpso is (237, 41, x, x, ..., x), and the corresponding binary string after encoding is 01101001001x... X (this example only gives the assignment of the first 18 bits in the binary string, and only involves the first and second dimensions).

In this process, the structure created is continuous. However, a grid generated by one father may prevent the next generation of another father. For example, in Fig. 3.1e, since the B grid generated nearby is earlier than the C grid, when the upper grid of B grid is set, although it is also the left grid of C grid, it prevents it from being set again. Due to blocking, in almost all processes, only less than 80 numbers are read when the stop condition occurs—although some grids are still “empty”. Therefore, using 100 binary strings wastes almost the last 20 or more binary digits. This means that a 90 binary string is enough to create a complete structure.

As an example, our target is a multiband monopole antenna, so the feeder should be connected to the antenna. To ensure this feature, after creating the above structure, all grids “0” are changed to grid “null”, and then the two grids in the middle of the bottom line of the design area are initialized to “1”. Then set the other meshes to “1” or “0” according to the above method. Considering that the connection between antenna and feeder does not need too many grids, only the last 27 bits of binary string are used to guide the generation of this connection.

3.2.2 A-CLPSO Instead of GA for the New Grid Setting

In the traditional optimization process of antenna gridding, each grid is mapped to a binary number, “0” or “1”, so antenna optimization becomes optimization of binary string. Obviously, for two binary strings with different numbers, the generated two have different structures, but will have the same number of different grids. According to experience, in most cases, if only a few of the binary numbers are different, these different grid structures will not lead to too much difference in the reflection coefficient of the antenna. Therefore, the traditional antenna grid optimization is naturally closer to the genetic algorithm. It is through the genetic and mutation that the antenna structure changes slightly step by step, and finally converges to the best result. From the perspective of fitness function, when a binary code changes, the

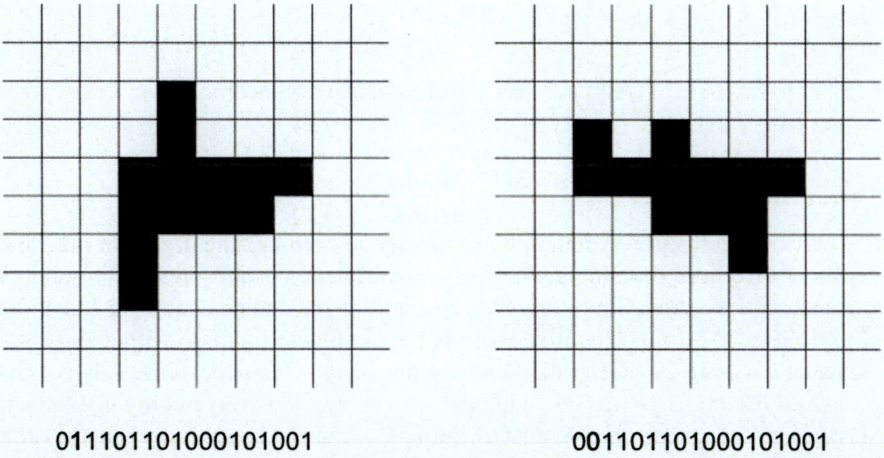

Fig. 3.2 Generated structures from two different binary string samples. Figure reproduced with permission from Ref. [15], © 2009 IEEE

corresponding fitness value usually only changes a little, and changes smoothly in the search area. In many literatures about genetic algorithm, we can see that genetic algorithm is suitable for solving this kind of problems.

In our new antenna gridding optimization method, although it is still based on binary string optimization, according to the grid assignment principle and steps mentioned in the previous section, the grid is not firmly connected to a certain number. In this way, the discontinuity in the process of antenna structure generation is avoided, and the order of mesh generation is determined by the generated mesh. The number change in the binary string will cause the grid assignment to change, and also affect the generation order of other grids to be assigned around it. For example, when the binary string "011011001001" in Fig. 3.1 is changed to "001101000101010001", the structure as shown in Fig. 3.2 is generated, which is very different. The results show that as long as a number in the binary string changes, it may cause huge differences in structure. Therefore, the corresponding fitness function extremum optimization problem is more like a multi-modal solution problem in a volatile search region. This chapter is based on CLPSO algorithm and improves it to a-clpso algorithm to achieve the optimal performance of multimodal problems.

3.2.3 The Fitness Function

It is the key to find the suitable fitness function in practical optimization problems. For general antenna optimization, designers pay great attention to the frequency band of | S11 | < −10 dB, so the fitness function of early antenna design optimization is set to only focus on the type of return loss satisfying the condition in the required band [14], for example:

$$f = \sum S(i) \quad i \in \text{the bands we concern} \tag{3.1}$$

where $S(dB) = \begin{cases} S_{11}(dB) + 10dB & S_{11}(dB) > -10dB \\ 0 & S_{11}(dB) < -10dB \end{cases}$.

Obviously, this fitness function can be used to optimize the design of broadband antenna. However, this single objective fitness function is not suitable for multiband antenna optimization. Suppose to design a multi band antenna covering 1–1.1 GHz, 2–2.1 GHz, 3–3.1 GHz, so it is easier to set the fitness function with a super band antenna covering 2–4 GHz than with a multi band antenna covering 1.1–1.2 GHz, 2.1–2.2 GHz and 3.1–3.2 GHz with high efficiency. However, empirically, the latter is more accurate and valuable than the former, because it is easier to change a certain resonant frequency by adjusting some parts of the antenna. Therefore, the traditional single target fitness function can not correctly evaluate the prototype of multi band antenna. Here, we propose a new multi band antenna fitness function, which involves bandwidth and frequency offset

$$f = \sum_{i=1}^{N} F(i) \tag{3.2}$$

$$F(i) = \begin{cases} \frac{|f_{mid}(i) - f_0(i)| + 1\ \text{MHz}}{bandwidth(i)} & \text{can find a below - 10 dB band in}\left[f_{\inf}(i), f_{\sup}(i)\right] \\ S & \text{cannot find a below - 10 dB band in}\left[f_{\inf}(i), f_{\sup}(i)\right] \end{cases} \tag{3.3}$$

N is the number of frequency bands that we need to design the antenna, $f_{mid}(i)$ is the central frequency of the ith found band in $[f_{\inf}(i), f_{\sup}(i)]$, $f_0(i)$ is the central frequency of the ith desired band, $bandwidth(i)$ is the -10 dB impedance bandwidth of the ith found band, $[f_{\inf}(i), f_{\sup}(i)]$ is the ith search range which is wider than the ith desired bandwidth, S is a constant, If the antenna has no satisfied frequency band, the value of this constant S is given relatively large, thus giving a bad evaluation.

To the civil wireless transmission, the expected multi-band monopole antenna needs to cover 1.88–2.025 GHz, 2.4–2.5 GHz, 3.4–3.6 GHz and 5.15–5.85 GHz. Here, $[f_{\inf}(1 \ldots 4), f_{\sup}(1 \ldots 4)]$ are set as [1.7, 2.2], [2.2, 2.7], [3, 4], [4.5, 6] and S is assigned as 50—it is large enough. Even if the frequency offset is 500 MHz and the available bandwidth is only 10 MHz, the value of $F(i)$ is still less than S.

In the coarse model design of the antenna, the design area is 17 mm × 26 mm, evenly divided into 10 × 10 squares. Since the dielectric constant of the substrate is 2.65 and the thickness is 1 mm, the feeder width of the antenna port is set as 2.8 mm, corresponding to the characteristic impedance of 50 Ω. The feeder length is 32 mm and the floor size is 29.6 mm × 26 mm. In order to illustrate the difference between the two fitness functions, we use 30 particles and a-clpso algorithm to solve the two fitness functions. The figure of the solution results of the two fitness functions and the simulated reflection coefficient curve of the corresponding antenna are shown in Figs. 3.3, 3.4 and 3.5 respectively. It can be seen from the results that the antenna structure optimized by the traditional objective function is more like an ultra wideband antenna and can not cover the lowest frequency band, while the antenna structure optimized by the new objective function obviously has multiple frequency bands, and each frequency band is close to the required design frequency band. It also shows that the new fitness function can be used in the next step of antenna design.

Fig. 3.3 Structure of the optimized multi-band monopole antenna after rough design using traditional fitness function (black: copper, white: substrate, grey: ground). Figure reproduced with permission from Ref. [15], © 2009 IEEE

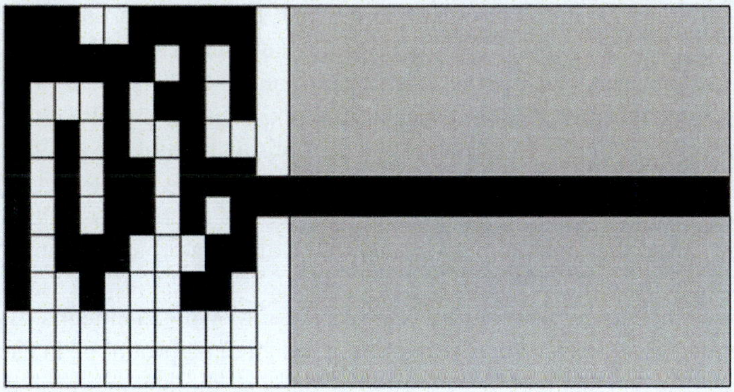

Fig. 3.4 Structure of the optimized multi-band monopole antenna after rough design using the new fitness function (black: copper, white: substrate, grey: ground). Figure reproduced with permission from Ref. [15], © 2009 IEEE

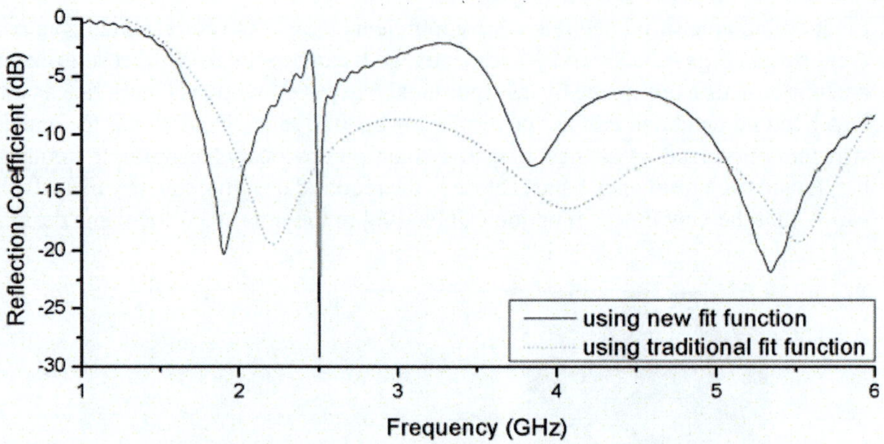

Fig. 3.5 The simulated reflection coefficient curves of the two structures with different fitness functions. Figure reproduced with permission from Ref. [15], © 2009 IEEE

3.2.4 Prototype Modification

Although the optimization results of antenna coarse model design successfully obtain the characteristics of multi band coverage, the digital discrete structure based on binary system also brings some practical problems, such as the high requirement of PCB manufacturing accuracy and the difficulty of structure adjustment. In addition, the reflection coefficient of the prototype antenna with binary digital discrete structure has not fully met the initial design requirements. The main reason is 10 × The mesh is rough and can not express the details of the antenna. Although further increasing the number of grids can improve the calculation accuracy, it will also lead to the increase of the number of grids, especially the exponential growth of the calculation time. Therefore, in order to further refine the optimization of the antenna, we regard the antenna prototype as a continuous structure, which is also the first adjustment of generating smooth structure and generating rules. Of course, this change is only an improvement of the reflection coefficient of the antenna prototype, not a big change of the reflection coefficient curve.

The current density distribution is the key of the resonant antenna. By analyzing the simulated current density distribution of the antenna prototype at each expected frequency, we can see that some "strange" parts of the antenna prototype may be modulated into "regular" parts. Of course, the modification of each part follows the principle mentioned above, that is, the curve of reflection coefficient is only a small modification, not a big change. Through our modification, the prototype is first adjusted to Fig. 3.6a, then Fig. 3.6b, and finally Fig. 3.6c. This modification process is simple and easy. The simulated reflection coefficient curve of the corresponding antenna is shown in Fig. 3.7. The results show that the characteristics of the antenna

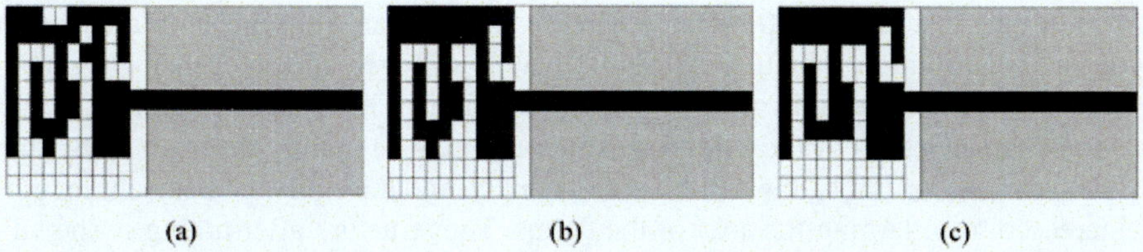

Fig. 3.6 The 3 structures in changing process of the prototype. Figure reproduced with permission from Ref. [15], © 2009 IEEE

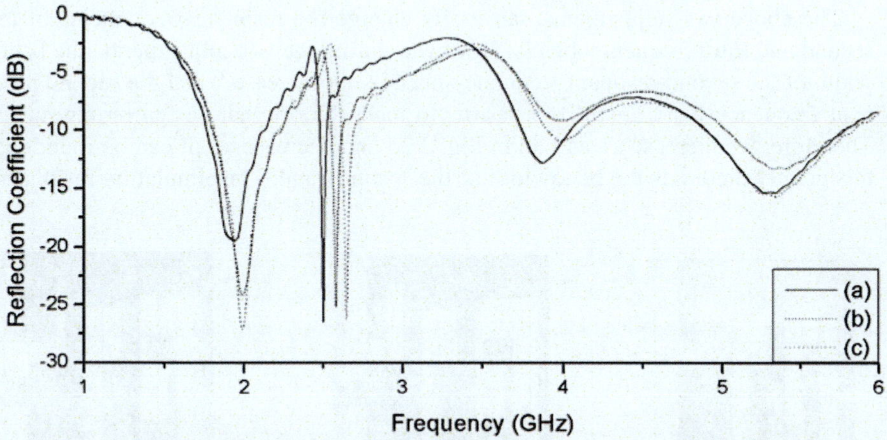

Fig. 3.7 The simulated reflection coefficient curves of the 3 structures in Fig. 3.6. Figure reproduced with permission from Ref. [15], © 2009 IEEE

do not change much after modifying from Fig. 3.4, 3.5 and 3.6c, which is still called the prototype structure.

3.2.5 Further Adjustment

The original structure antenna adjusted according to the above steps can cover four frequency bands. However, the bandwidth and resonant frequency of some frequency bands still can not meet the requirements, and need to be further fine tuned. It can be seen from the simulation that changing the shape of the left side of part a and part B in Fig. 3.8a, b can increase the equivalent current length of the second frequency band and move the resonance frequency point to the low frequency. Figure 3.9 shows the simulated reflection coefficient of the antenna prototype after changing the above two parts. Obviously, the resonant frequency of the second band is reduced from 2.50 to 2.37 GHz.

For the third resonant frequency band, it can be seen from the simulated current density distribution that the distribution with stronger current density is mainly concentrated on the right side of parts B and D in Fig. 3.8a. Therefore, the radiation characteristics of the third resonant band may be mainly determined by these two parts. Accordingly, the d part is tuned firstly, and the equivalent current length is increased by enlarging the area of the d part. The structure after tuning is shown in Fig. 3.8c, and the simulated reflection coefficient of the corresponding antenna prototype is shown in Fig. 3.10. Obviously, this adjustment makes the resonant frequency of the third band move from 3.82 to 3.52 GHz, but has little effect on the bandwidth of the third resonant frequency.

The above two adjustments can easily change the resonant frequencies of the second and third frequency bands. However, during the two adjustments, the bandwidth of the second resonant frequency band is not increased, and the second resonant frequency band is still too narrow to meet the original design requirements. Therefore, we increase a1 and a2 in Fig. 3.8d and decrease a3. It can be found that this greatly increases the bandwidth of the second band. The simulation results are

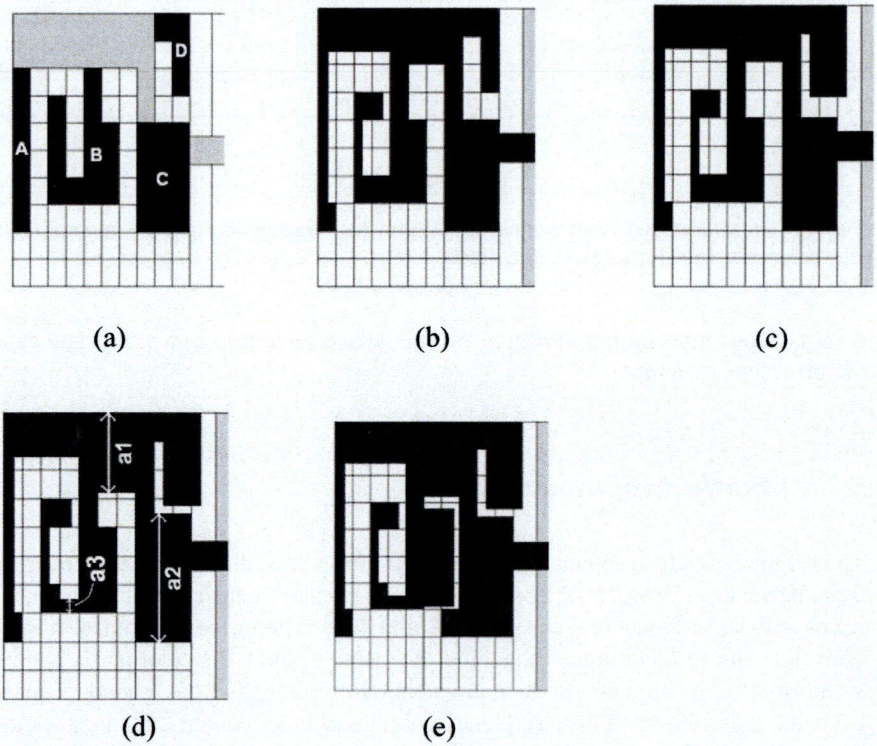

Fig. 3.8 The raw structure and 4 modified structures in further adjustment. Figure reproduced with permission from Ref. [15], © 2009 IEEE

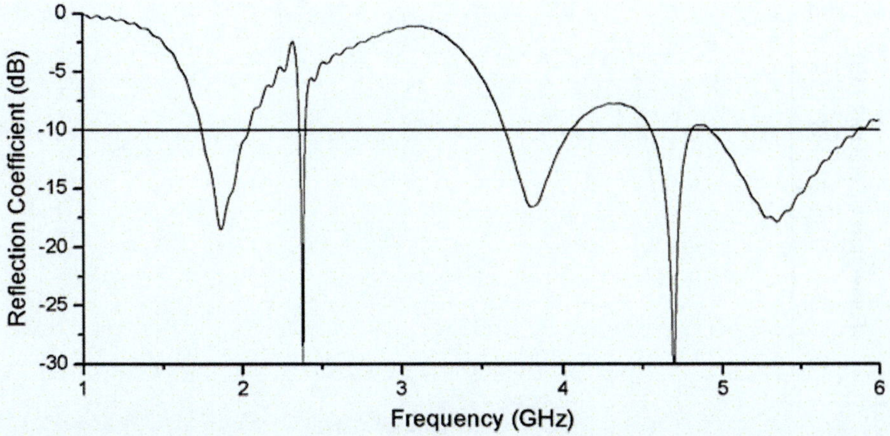

Fig. 3.9 The simulated reflection coefficient of Fig. 3.8b. Figure reproduced with permission from Ref. [15], © 2009 IEEE

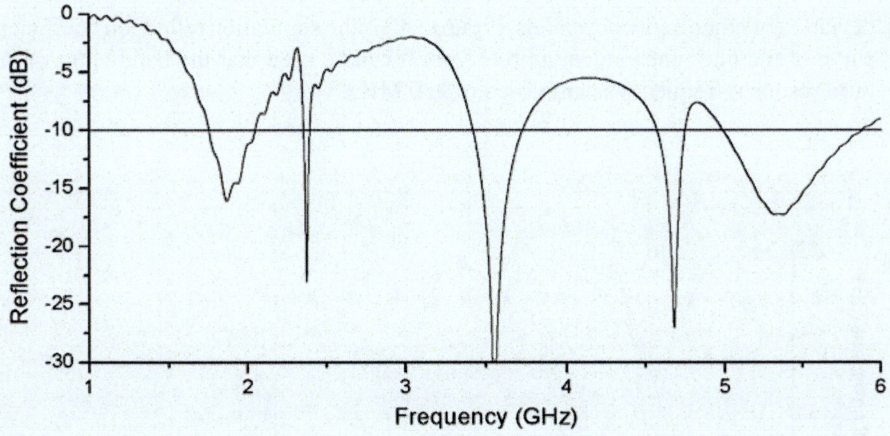

Fig. 3.10 The simulated reflection coefficient of Fig. 3.8c. Figure reproduced with permission from Ref. [15], © 2009 IEEE

shown in Fig. 3.11. However, this adjustment process has seriously deteriorated the bandwidth of the third band.

From the previous research, it can be seen that the right side of Part B and Part D may affect the characteristics of the third frequency band, and Part D mainly determines the position of the resonant frequency, so the right side of Part B is tuned here. Simulation results show that increasing the area of B part can increase the bandwidth of the third band, but its resonant frequency position will deviate from the expected frequency. Furthermore, adding a small piece of metal at the bottom left of Part C can offset the deviation of resonant frequency, but it will lead to a slight

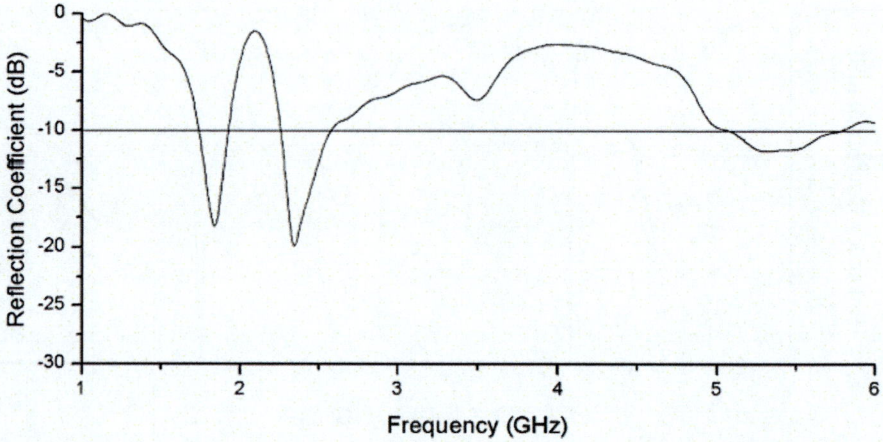

Fig. 3.11 The simulated reflection coefficient of Fig. 3.8d. Figure reproduced with permission from Ref. [15], © 2009 IEEE

deviation of other frequency bands. Figure 3.12 is the simulation reflection coefficient curve of the fine tuned antenna prototype. It can be seen that the bandwidth of the third resonant frequency is amplified to 250 MHz.

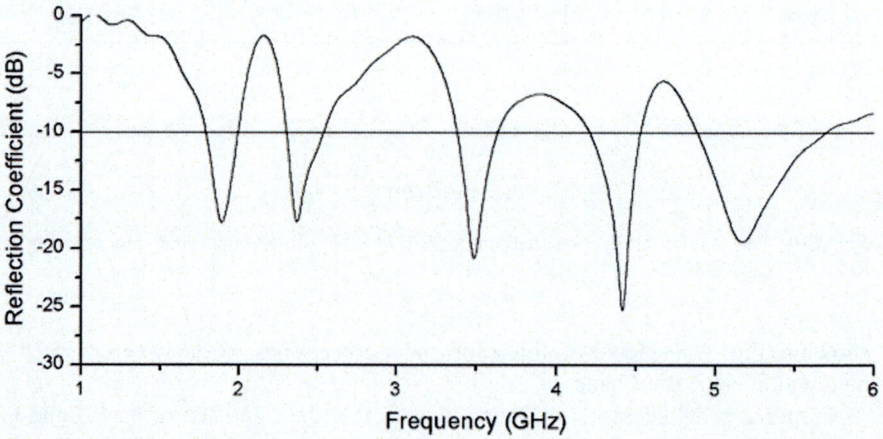

Fig. 3.12 The simulated reflection coefficient of Fig. 3.8e. Figure reproduced with permission from Ref. [15], © 2009 IEEE

3.2.6 Optimized Result

From the above adjustment and analysis process, it can be seen that through the adjustment of the original structure of Fig. 3.8a–e, some key parts of the antenna prototype have a significant impact on the characteristics of the antenna. However, these parts not only affect a single frequency band, only modifying one part to improve some characteristics will deteriorate the other. Therefore, it is necessary to introduce optimization algorithm to achieve effective coverage of multiple frequency bands through collaborative optimization of various parts. In this section, 12 parameters (such as X1 to X12 in Fig. 3.13) are set on the adjusted structure to characterize the structure of the antenna prototype, and the goal of multi band coverage is achieved through the overall optimization of these parameters, while other size parameters that have little contribution to the characteristics are set as constants (A1-5, B1-5, C1-2, D1-3, E1-4 in Fig. 3.13). In order to avoid some parts overlapping when the parameters change, the final structure of the antenna prototype has some minor changes to the shape of the original structure after adjustment. Similarly, we use a-clpso algorithm with 30 particles to optimize these 12 parameters. The final values of all dimensions

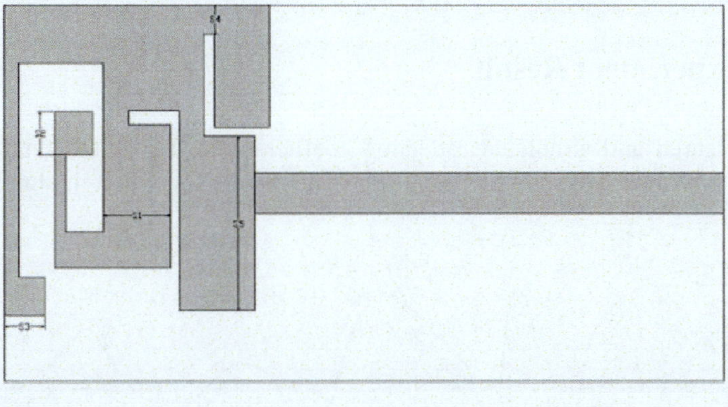

Fig. 3.13 Final structure of the multi-band monopole antenna. Figure reproduced with permission from Ref. [15], © 2009 IEEE

Table 3.1 The values of the 12 parameters (all are rounded to the nearest 0.1 mm) and other sizes

X1 = 4.8	X7 = 1.2	A1 = 6.8	B1 = 3.4	C1 = 18.8	D1 = 8.5	E1 = 6.8
X2 = 6.8	X8 = 3.6	A2 = 7.3	B2 = 1.6	C2 = 6.8	D2 = 0.6	E2 = 10.9
X3 = 2.5	X9 = 2.1	A3 = 1.7	B3 = 5.2		D3 = 4.4	E3 = 3.4
X4 = 2.7	X10 = 3.7	A4 = 1.0	B4 = 1.7			E4 = 1.7
X5 = 2.6	X11 = 11.7	A5 = 0.5	B5 = 2.6			
X6 = 2.7	X12 = 6.8					

Unit (mm)

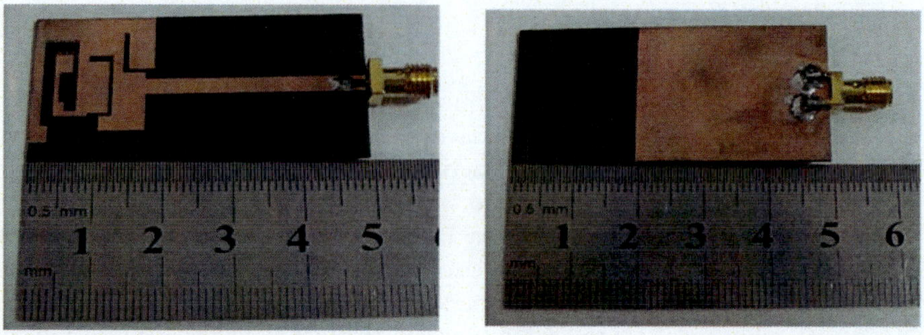

(a) (b)

Fig. 3.14 Photograph of the proposed multi-band monopole antenna. **a** geometry, **b** ground. Figure reproduced with permission from Ref. [15], © 2009 IEEE

of the optimized antenna prototype are shown in Table 3.1, and the real photos of the optimized multi band antenna prototype are shown in Fig. 3.14.

3.3 Experiment Result

The measured and simulated reflection coefficient curves of the final processed antenna prototype are shown in Fig. 3.15. The results show that the final reflection

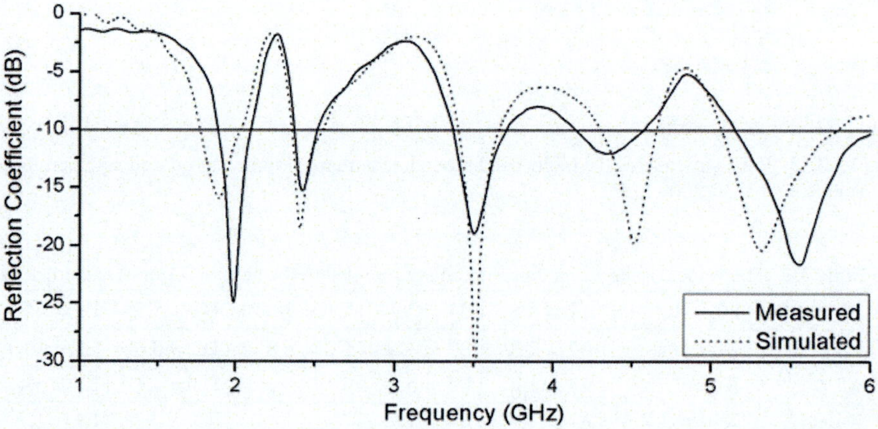

Fig. 3.15 The measured and simulated reflection coefficient of the final structure (the five band with the measured reflection coefficient under −10 dB are **a** 1.88–2.10 GHz, **b** 2.37–2.52 GHz, **c** 3.37–3.71 GHz, **d** 4.14–4.58 GHz, **e** 5.14–6.00 GHz). Figure reproduced with permission from Ref. [15], © 2009 IEEE

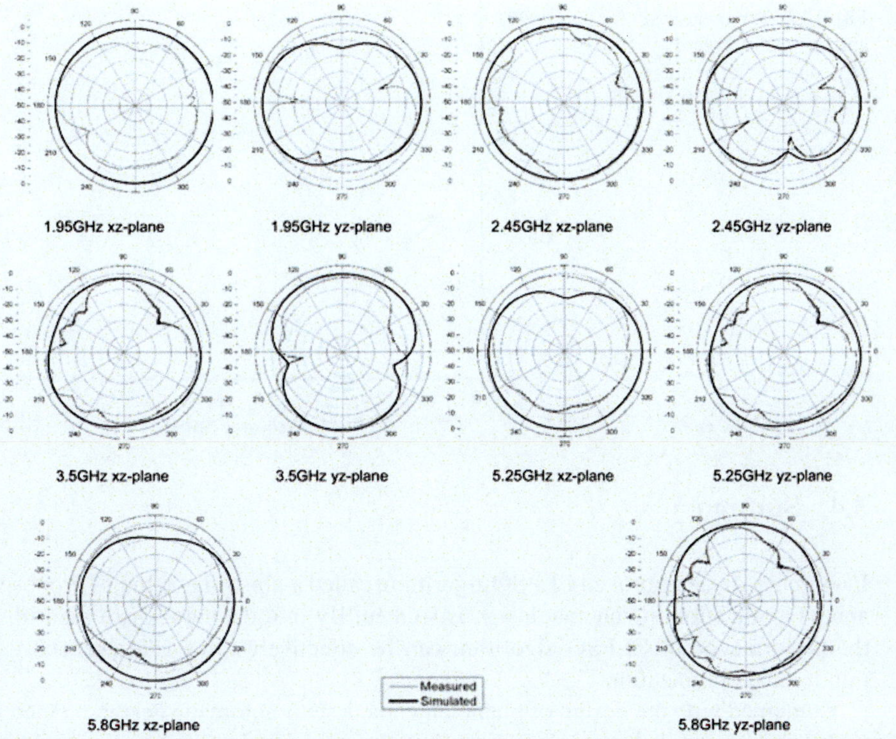

Fig. 3.16 Measured and simulated radiation patterns at different frequencies. Figure reproduced with permission from Ref. [15], © 2009 IEEE

coefficient of the antenna basically covers several required wireless communication frequency bands, meeting the initial design requirements.

The measured two-dimensional far-field radiation patterns of the antenna at different frequencies are shown in Fig. 3.16.

It can be seen from Fig. 3.16 that the measured radiation pattern is basically consistent with the simulated radiation pattern. At the desired frequency, the near omnidirectional radiation pattern in the xoz plane and the dipole like radiation pattern in the YOZ plane are measured. Since the radiation pattern is measured on the roof of a five story building, there may be some differences between the measured and simulated radiation patterns in this semi open space and other surrounding buildings. The measured antenna gain curve is shown in Fig. 3.17. It can be seen that the antenna has a high measured gain in the high frequency band. The gain range is 4.5–5.3 dB at 3.5, 5.25 and 5.8 GHz.

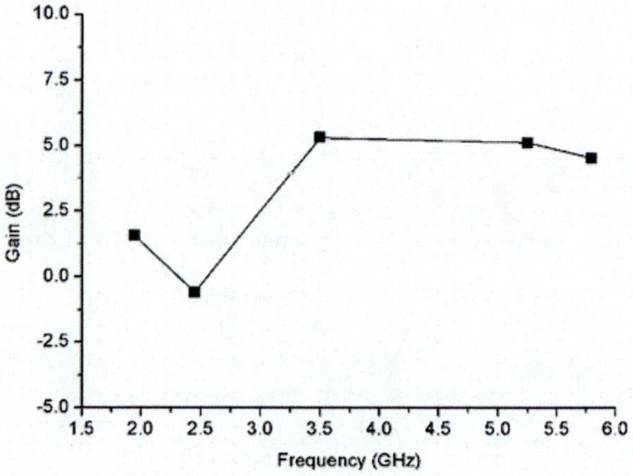

Fig. 3.17 The measured antenna gains. Figure reproduced with permission from Ref. [15], © 2009 IEEE

3.4 Summary

In this chapter, an improved CLPSO algorithm, called a-clpso algorithm, is proposed, and its application in antenna design is studied. By introducing a-clpso algorithm, the performance of CLPSO algorithm can be effectively improved, especially in multimodal optimization.

Combined with the design of planar antenna, a semi-automatic design method of multi band monopole antenna is proposed. In antenna optimization, a new grid setting method is proposed. A-clpso algorithm is used to optimize the rough model design problem with multi-modal characteristics. Due to the introduction of a new fitness function in the optimization, it is more suitable for the design of multi band antenna. After rough model design and precision model design, the antenna prototype is optimized and tested. The experimental results show that the multi frequency antenna can successfully cover the required frequency bands of TD-SCDMA, WLAN and WiMAX, and the a-clpso algorithm and its application in semi-automatic antenna design are very effective.

References

1. M. Pokorn, J. Horak, Design and global multi-objective optimization of planar tri-band antenna, in Radioelektronika, 2007. 17th International Conference, 2007, pp. 1–5
2. M. Ding, Optimal design of ultra wideband antennas using a mixed model of 2-D genetic algorithm and finite-difference time-domain. Microw. Opt. Technol. Lett. **49**(12), 3177–3180 (2007)
3. G. Kim, Optimization of UHF RFID tag antennas using a genetic algorithm, in Antennas and Propagation Society International Symposium, 2006, pp. 2087–2090
4. J. Kennedy, R. Eberhart, Particle swarm optimization, in IEEE International Conference on Neural Networks, Perth, Australia, vol. 4, 1995, pp. 1942–1948

5. F. van den Bergh, A.P. Engelbrecht, A cooperative approach to particle swarm optimization. IEEE Trans. Evol. Comput. **8**(3), 225–239 (2004)
6. S. Cui, D.S. Weile, Application of parallel particle swarm optimization scheme to the design of electromagnetic absorbers. IEEE Trans. Antennas Propag. **53**(11), 3616–3624 (2005)
7. S.M. Mikki, A.A. Kishk, Quantum particle swarm optimization for electromagnetics. IEEE Trans. Antennas Propag. **54**(10), 2764–2775 (2006)
8. S. Genovesi, A. Monorchio, R. Mittra, G. Manara, A sub-boundary approach for enhanced particle swarm optimization and its application to the design of artificial magnetic conductors. IEEE Trans. Antennas Propag. **55**(33), 766–770 (2007)
9. T. Huang, A.S. Mohan, A microparticle swarm optimizer for the reconstruction of microwave images. IEEE Trans. Antennas Propag. **55**(3), 568–576 (2007)
10. T. Huang, A.S. Mohan, A hybrid boundary condition for robust particle swarm optimization. IEEE Antennas Wireless Propag. Lett. **4**, 112–117 (2005)
11. S. Selleri, M. Mussetta, P. Pirinoli, R.E. Zich, L. Matekovits, Differentiated meta-PSO methods for array optimization. IEEE Trans. Evol. Comput. **56**(1), 67–75 (2008)
12. J.J. Liang, Comprehensive learning particle swarm optimizer for global optimization of multimodal functions. IEEE Trans. Evol. Comput. **10**(3), 281–295 (2006)
13. A. Ratnaweera, S.K. Halgamuge, H.C. Watson, Self-organizing hierarchical particle swarm optimizer with time-varying acceleration coefficients. IEEE Trans. Evol. Comput. **8**(3), 240–255 (2004)
14. D.H. Wolpert, W.G. Macready, No free lunch theorems for optimization. IEEE Trans. Evol. Comput. **1**(1), 67–82 (1997)
15. H. Wu, J. Geng, R. Jin, J. Qiu, W. Liu, J. Chen, S. Liu, An improved comprehensive learning particle swarm optimization and its application to the semi-automatic design of antennas. IEEE Trans. Antennas Propag. **57**(10), 3018–3028 (2009)

Chapter 4
Design Low Profile RFID Tag Antenna on Metal with Binary Coding Optimization Method

4.1 Introduction

Radio-frequency identification (RFID) technology has become an indispensable technology in social life. In people's daily life, traveling, dining, shopping and so on need RFID technology for identification, even payment. Radio frequency identification (RFID) technology has been greatly developed in recent years. Because the contactless data transmission between reader and tag is more flexible than the mechanical contact used in automatic identification system [1, 2], it has become an important method in industry and commerce. Especially in recent years, with the rapid development of logistics and warehousing, RFID technology has become the key to the transportation and storage of industrial products and commodities, which can help people clearly and automatically identify goods, count goods, record goods, and facilitate the collection and transmission of data.

At the beginning of the twentieth century, with the birth of radar, people began to use the reflection of radio waves to detect and locate objects. In 1948, Harry stockman published a paper entitled "communication through reflected power", which is an early exploratory work on RFID. The 1950s is the era of RFID technology exploration, some of which include the "identify friend or enemy" (IFF) remote transponder system for military aircraft and the radio transmission system with adjustable passive responder.

In 1964, Harrington studied the electromagnetic theory related to RFID in his papers such as "load scattering theory". At the same time, many researchers are also carrying out RFID related inventions, such as "remote activated RF power supply equipment", "passive data transmission" and "technology using radar echo".

In the 1960s, the first commercial RFID appeared. Some international companies have developed the well known electronic article surveillance (EAS) equipment to prevent articles from being removed from the control area without authorization [3]. In the EAS system, the EAS tag will interact with the electromagnetic field located in the controlled area, so as to detect the unauthorized moving objects and take measures such as warning. The EAS tag can also be used to authorize the movement

J. Geng and R. Jin, *Antenna Optimization and Design Based on Binary Coding*, Modern Antenna, https://doi.org/10.1007/978-981-16-7965-0_4

of goods. The device authorized to disable the label is equipped with a coil. When it is powered on, the magnetic field generated can make the label invalid, that is, it will no longer respond to the incident energy of the system.

In the 1970s, more institutions and laboratories joined the research work of RFID, and made significant progress. For example, Los Alamos National Laboratory integrates RFID tags into employees' certificates to automatically identify employees and restrict access to safe areas [4]. In addition, it also applies RFID to animal tracking. Large companies are also developing RFID technology. For example, the port authorities in New York and New Zealand have tested the systems developed by general electric and Philips. In 1973, RFID was allowed to be used in the development of electronic vehicle identification in the United States.

In the 1980s, RFID technology has been widely used. RFID has been used in transportation and personnel access management in the United States, as well as in animal tracking and industrial short-range systems in Europe. Toll roads in many countries are equipped with RFID. At this time, CMOS integrated circuits and discrete components are used in the production of tags. At the same time, EEPROM also makes the large-scale production of the same tag possible. These make RFID tags further reduce, and the constraint of antenna size is becoming more and more important.

In the 1990s, the electronic toll collection system in the United States deployed RFID on a large scale and installed tags on more than 3 million cars. Railway toll has gradually appeared in various countries around the world. With the successful use of electronic toll collection system, tags are also used across business areas, such as parking lot, community and campus access.

In the twenty-first century, the application of RFID is more common, including but not limited to the prevention of theft, railway charges, automatic parking, distribution of goods. And its international standards in different application fields are gradually established.

As China has the largest electronic market in the world, the rapid development of express industry and logistics industry brought by the rise of e-commerce also provides a new direction for the application of RFID technology. The application of RFID can greatly reduce the consumption of manpower and improve the sorting efficiency of express delivery. With the 5 g construction and the maturity of blockchain technology, the express Internet of things with electronic tags as the core has also developed rapidly.

RFID system consists of reader and tag antenna. Tag antenna is an important part of signal exchange in RFID system. Microwave tag is composed of at least CMOS integrated circuit and antenna, and can be pasted on the object. This makes the circuit size smaller, cost lower and reliability improved. At this time, the tag size is limited by the antenna, and the appropriate antenna design becomes a big challenge in the field of RFID technology [3]. There are many kinds of tag antennas based on dipole or circle. However, the performance of tag antenna is largely affected by its surrounding environment. When the tag antenna is connected to a metal surface, such as a refrigerator or container, it can hardly send and receive signals. This is mainly because the mirror current on the metal surface is opposite to that of the tag antenna, and the effective signal is cancelled. One way to solve this problem is to patch mushroom EBG structure on the metal surface, but the structure is complex and the cost is high [5].

In order to solve this problem, an antenna structure based on particle swarm optimization (PSO) is proposed in this chapter, which is hollow engraving patch antenna part and hollow middle ground layer [6]. This structure can work on the metal grounding of China's RFID frequency band 840–845 MHz, and can also cover multiple frequency bands including RFID and GPS.

4.2 The Basic Binary-Coding Structure for Antenna

Although there are many kinds of tag antennas in RFID system, most of them are electric dipole type. When this kind of electric dipole antenna is pasted on the metal surface, it is difficult to achieve normal directional radiation due to the reverse mirror current. However, metal clad items such as containers and refrigerators are in great need of tag antennas. In order to solve this problem, we build a multi-layer tag antenna with metal floor according to the method in Chap. 2.

The tag antenna is composed of three layers of metal and two layers of dielectric, as shown in Fig. 4.1, which are the top layer binary coding patch, dielectric layer 1, the second layer binary coding patch, dielectric layer 2 and the floor. The coding layer is divided into 10 * 10 grids.

In essence, the top metal is a patch with many branches, and the distribution of current can stimulate the resonance of different frequency modes. The second layer of metal can exist as a ground, but it is really in the lowest layer. Therefore, the second layer of metal is more of a disturbance effect. On the one hand, it can change the overall equivalent current distribution through local reflection; on the other hand, it can destroy the mirror image of the floor to the top current, so that the original current and mirror current of the upper and lower layers have a certain radiation in the direction upward, At the same time, the radiation in the horizontal direction is offset. Of course, the first and second layer of grid metal also has an important impact on the impedance matching.

4.3 Implement of A-CLPSO and Generation of the Patch Antenna with Carved Middle Ground

The tag antenna of RFID is designed by using the coding and grid generation method in Chap. 3 and A-CLPSO algorithm. Similar to Fig. 4.1, the top patch layer and the middle floor layer are divided into 10 * 10 grids. A-CLPSO algorithm [7] introduces 20 particles, each particle has 10 dimensions. Each mesh represents one dimension of the particles. If the parameter of a certain size is 0, it means that the corresponding grid of antenna engraving part or engraving ground does not exist. If the parameter

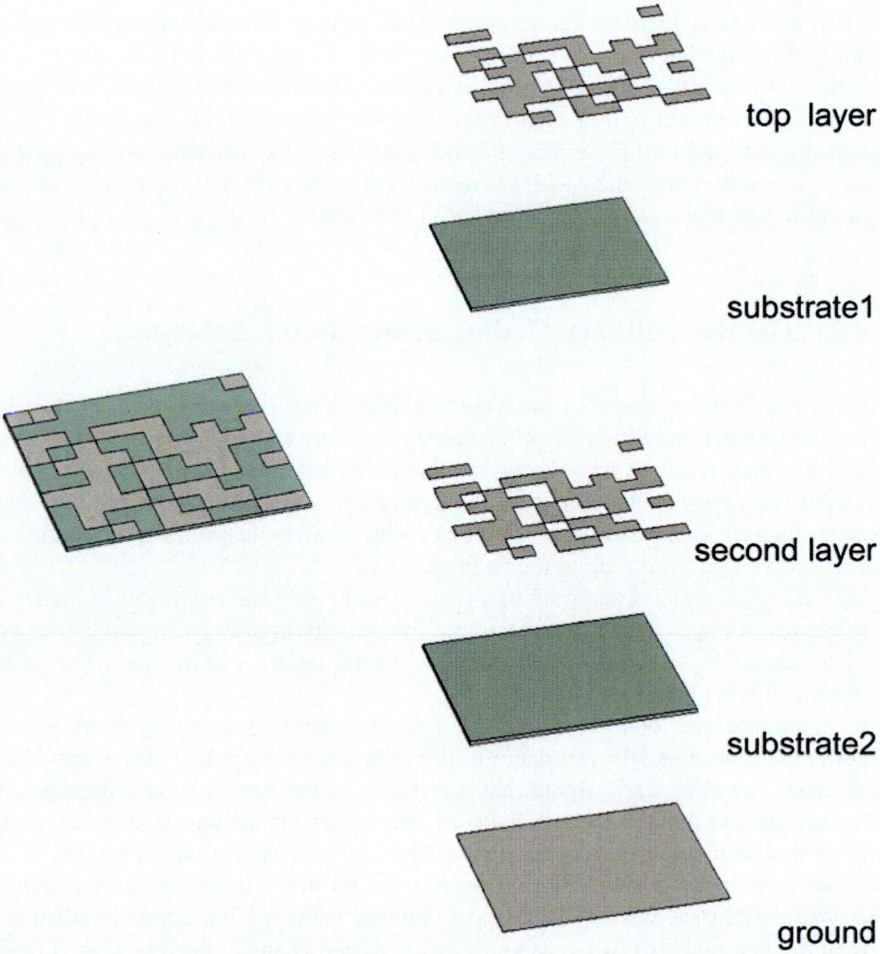

Fig. 4.1 The basic binary coding structure with two metal layer of rectangular mesh, two layers substrate and ground

is 1, the grid is part of the antenna or the ground. We need a total of 200 bits to control all the grids. The algorithm generates patch antennas with different shapes and carves the ground by changing these bits.

Due to the inaccuracy of simulation caused by the "point contact" in antenna design, we use the continuous growth method introduced in Chap. 3 to form the metal mesh for the upper patch, and we use the engraving hole method to generate the middle floor.

Fig. 4.2 Structure of the middle ground

4.3.1 *Parameters of the Antenna Design*

Chinese RFID frequency is 840–845 MHz, 920–925 MHz. In this RFID antenna design, we simulate the bandwidth of these two kinds of UHF. The design area of RFID tag antenna is 50 mm * 50 mm. It is divided into 10 × 10 grids. 50 Ω microstrip line input. The thickness of the two dielectric plates is 2 mm. Dielectric constant εr = 4.4.

The electromagnetic simulation software and a-clpso optimization package are combined to optimize the design according to the above basic binary coding grid structure. The optimized antenna structure is shown in Figs. 4.2, 4.3 and 4.4.

4.3.1.1 Result Analysis

Based on the optimized tag antenna in Fig. 4.4, the real manufactured antenna is in Fig. 4.5. The antenna's characters are simulated by the commercial electromagnetic software—CST, and the results are shown in Figs. 4.6 and 4.7.

In Fig. 4.6, the simulated |S11| is good in the frequency band 840–845 MHz and 920–925 MHz. In practice, the measured |S11| is good in 840–845 MHz, but the second band move right, which may be the result of the difference of dielectric constant of the two slabs.

Fig. 4.3 Structure of the patch

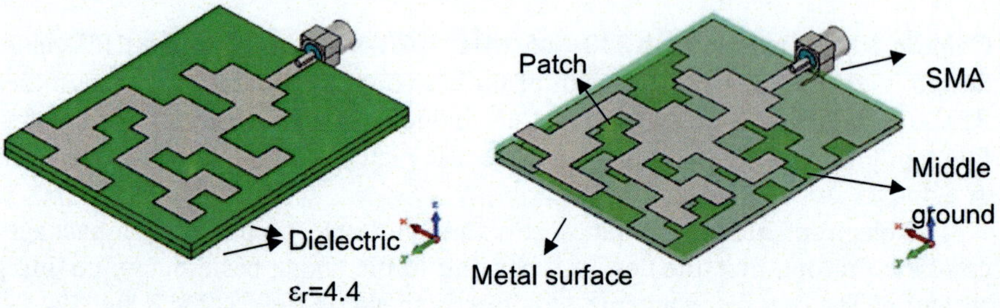

Fig. 4.4 Structure of RFID tag antenna

In Fig. 4.7, the simulated far field is directional in yz-plane, and almost omnidirectional in xy-plane. The simulated gain is 6.502 dBi.

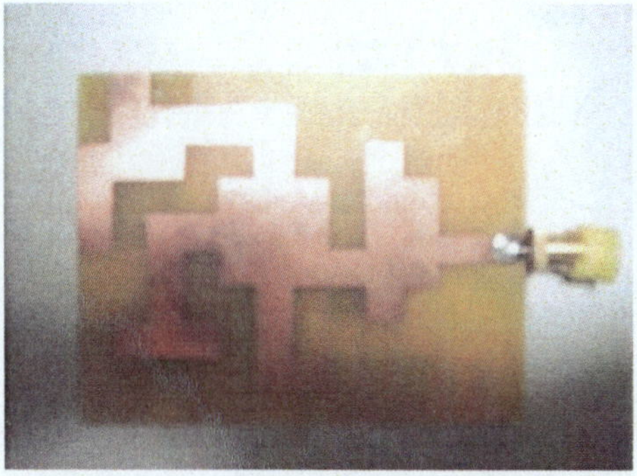

Fig. 4.5 The manufactured antenna. Figure reproduced with permission from Ref. [8], © 2010 IEEE

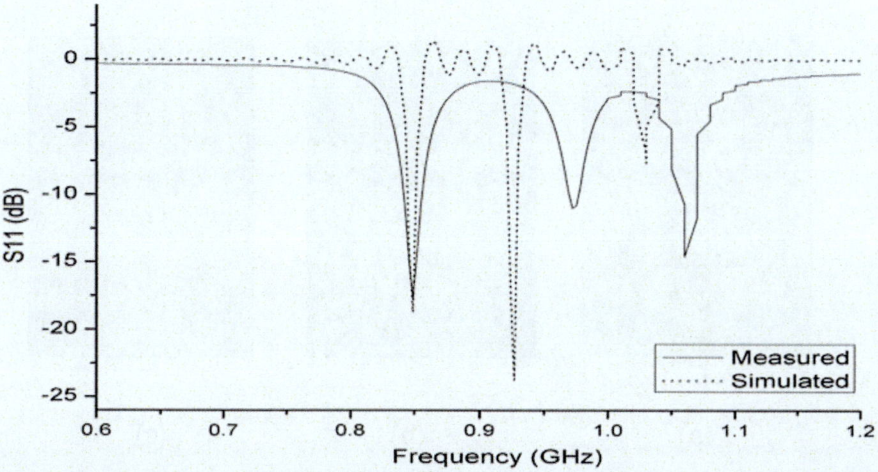

Fig. 4.6 Return loss |S11| of the tag antenna. Figure reproduced with permission from Ref. [8], © 2010 IEEE

4.4 RFID Tag Antenna is Optimized by Multi Objective Function and Other Algorithm

Furthermore, based on the binary coded grid structure in Fig. 4.1, three frequency bands (0.92–0.94 GHz, 1.24–1.25 GHz and 1.67–1.675 GHz) are set as the design objectives by using genetic algorithm, and the pattern is directionally radiated along the normal direction of the antenna, so that it can work on the metal surface. In this section, the optimization process of digging holes is adopted for the upper metal plate and intermediate transition ground. The structure of the optimized antenna is shown in Fig. 4.8.

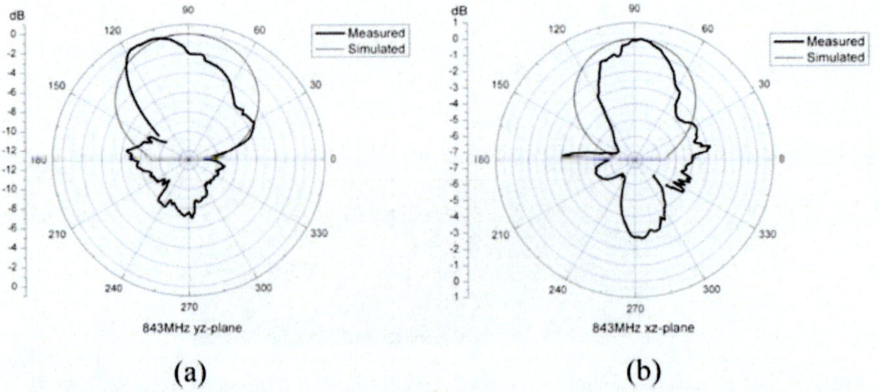

Fig. 4.7 The simulated and measured far field of the tag antenna: **a** yz-plane; **b** xz-plane. Figure reproduced with permission from Ref. [8], © 2010 IEEE

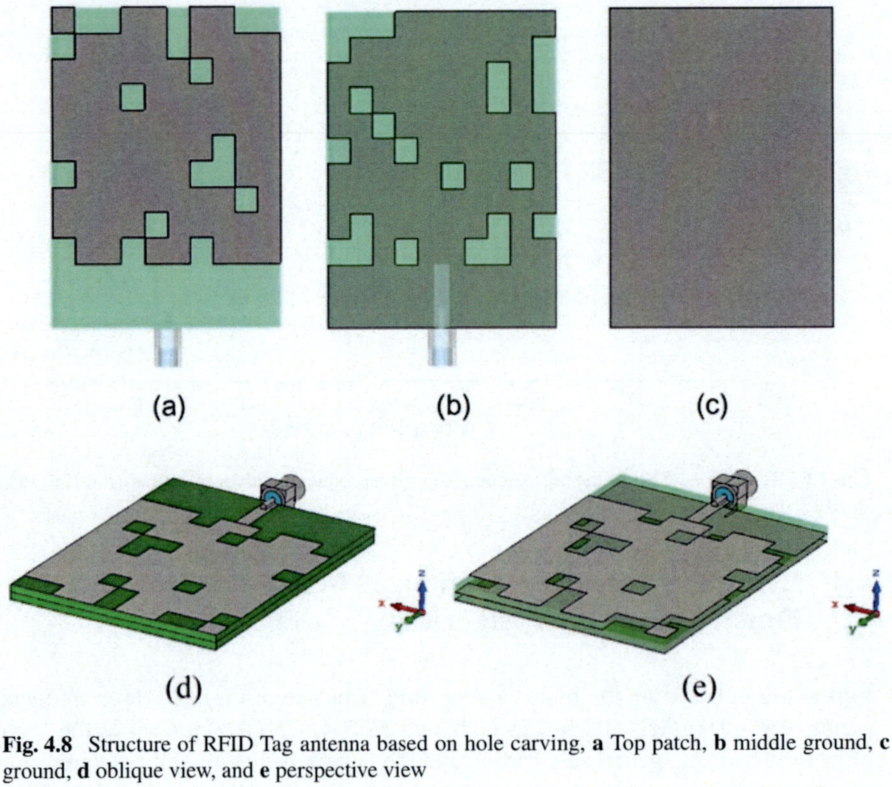

Fig. 4.8 Structure of RFID Tag antenna based on hole carving, **a** Top patch, **b** middle ground, **c** ground, **d** oblique view, and **e** perspective view

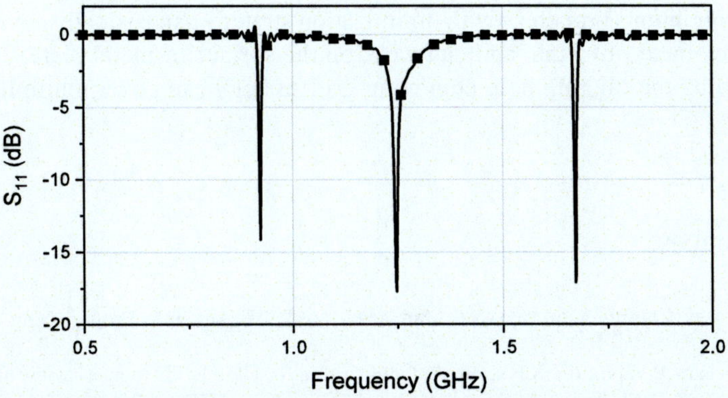

Fig. 4.9 The simulated return loss (S11) of the optimized RFID Tag antenna based on the hole carving. And there are three frequency bands (0.92–0.94 GHz, 1.24–1.25 GHz and 1.67–1.675 GHz) with |S11| ≤ 10 dB

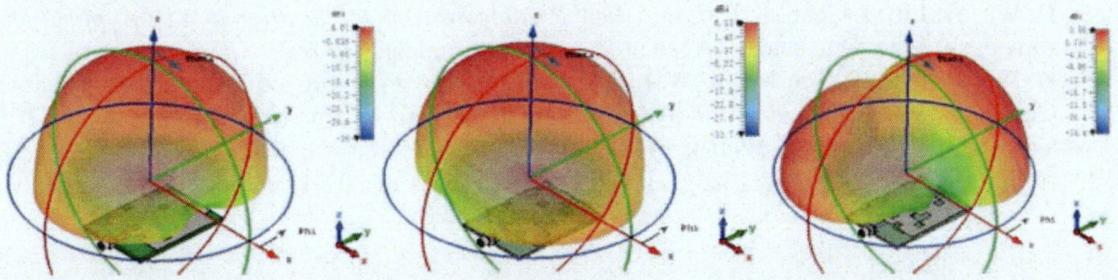

Fig. 4.10 The far field of the RFID Tag antenna at **a** 0.923 GHz and **b** 1.247 GHz with almost half sphere pattern, and **c** 1.673 GHz with split pattern

Different from the multi nodal structure of the upper segment, the optimization results in Fig. 4.8 are mainly patches, and the two-layer structure has several holes on the patch, which is similar to the open hole perturbation. The return loss curve of the optimized antenna structure is shown in Fig. 4.9, and the three target frequency bands basically meet the requirements. The pattern of the antenna is shown in Fig. 4.10. At 0.923 GHz, it is basically hemispherical and radiates upward, with a gain of 4dbi; At 1.247 GHz, the directional pattern is hemispherical and the gain is up to 6.3 dBi; The 1.647 GHz pattern bifurcates into two beams.

4.5 Summary

The double layer binary coding mesh model is introduced into the RFID tag antenna design, and different optimization results can be obtained by using different strategies and optimization methods, which can not only optimize the single objective antenna, but also optimize the multi-objective antenna. The RFID antenna obtained by the continuous branch growth optimization strategy can basically cover the RFID frequency band of China, and can work on the surface of metal plate. The antenna obtained by the digging hole engraving grid strategy can cover multiple frequency bands.

References

1. R. Bansal, Coming soon to a Wal-Mart near you. IEEE Antennas Propag. Mag. **45**, 105–106 (2003)
2. K.V.S. Rao, P.V. Nikitin, S.F. Lam, Antenna design for UHF RFID tags: a review and a practical application. IEEE Trans. Antennas Propag. **53**(12), 3870–3876 (2005)
3. J. Landt, The history of RFID. IEEE Potent. **24**(4), 8–11
4. R. Wantc, An introduction to RFID technology. IEEE Perv. Comput. **5**(1), 25–33
5. B. Gao, M.M.F. Yuen, Passive UHF RFID with ferrite electromagnetic band gap (EBG) material for metal objects tracking, in *Electronic Components and Technology Conference* (2008). *ECTC 2008. 58th*, pp. 1990–1994
6. H. Wu, W. Liu, J. Chen, S. Liu, in *A New Particle Swarm Optimization and Its Application to Antenna Design*. Outstanding Academic Papers by Students (2009)
7. H. Wu, J. Geng, R. Jin, J. Qiu, W. Liu, J. Chen, S. Liu, An improved comprehensive learning particle swarm optimization and its application to the semi-automatic design of antennas. IEEE Trans. Antennas Propag. **57**(10), 3018–3028 (2009)
8. X. Tao, J. Geng, R. Jin, X. Liang, H. Wu, RFID Tag Antenna for Use on Metal. iWAT

Chapter 5
Design Wideband Directional Antenna with Low Profile by Binary Coding Optimization Method

Equidistant helical antenna is a kind of broadband circularly polarized antenna, but it radiates bidirectional pattern. The conventional method is to put the equidistant spiral antenna on the floor and make it produce directional radiation through the reflection of the floor. The distance between the floor and the antenna is generally set at a quarter wavelength. But it will affect the bandwidth of spiral antenna, such as impedance bandwidth, axial ratio bandwidth and stable pattern bandwidth. This chapter firstly introduces the broadband balun [5, 6] used for spiral antenna feeding, and then based on the optimization model of microstrip antenna, the parasitic layer is rotationally symmetric to design the CP antenna with low profile. By optimizing the parameters of parasitic layer and spiral antenna at the same time, two kinds of antenna models are obtained. The directional low profile wideband circularly polarized helical antennas with different operating frequency bands are finally realized.

5.1 Design of Directional Broadband Circularly Polarized Helical Antenna Based on Involute Coaxial Line Feeding

5.1.1 Antenna Structure

The prototype structure of the broadband circularly polarized equiangular spiral antenna is shown in Fig. 5.1. The antenna has four metal layers and three layers of dielectric plates. The first layer of the metal is equidistant spiral antenna, the surface of the second layer of dielectric plate is attached with the first layer of parasitic patch, the front part of the third layer of dielectric plate is attached with the second layer of parasitic patch, and the rear part is attached with the floor. The dielectric constant of the three layers is the same ($\varepsilon_{r1} = \varepsilon_{r2} = \varepsilon_{r3} = 2.65$), and the thickness is 1 mm,

J. Geng and R. Jin, *Antenna Optimization and Design Based on Binary Coding*, Modern Antenna, https://doi.org/10.1007/978-981-16-7965-0_5

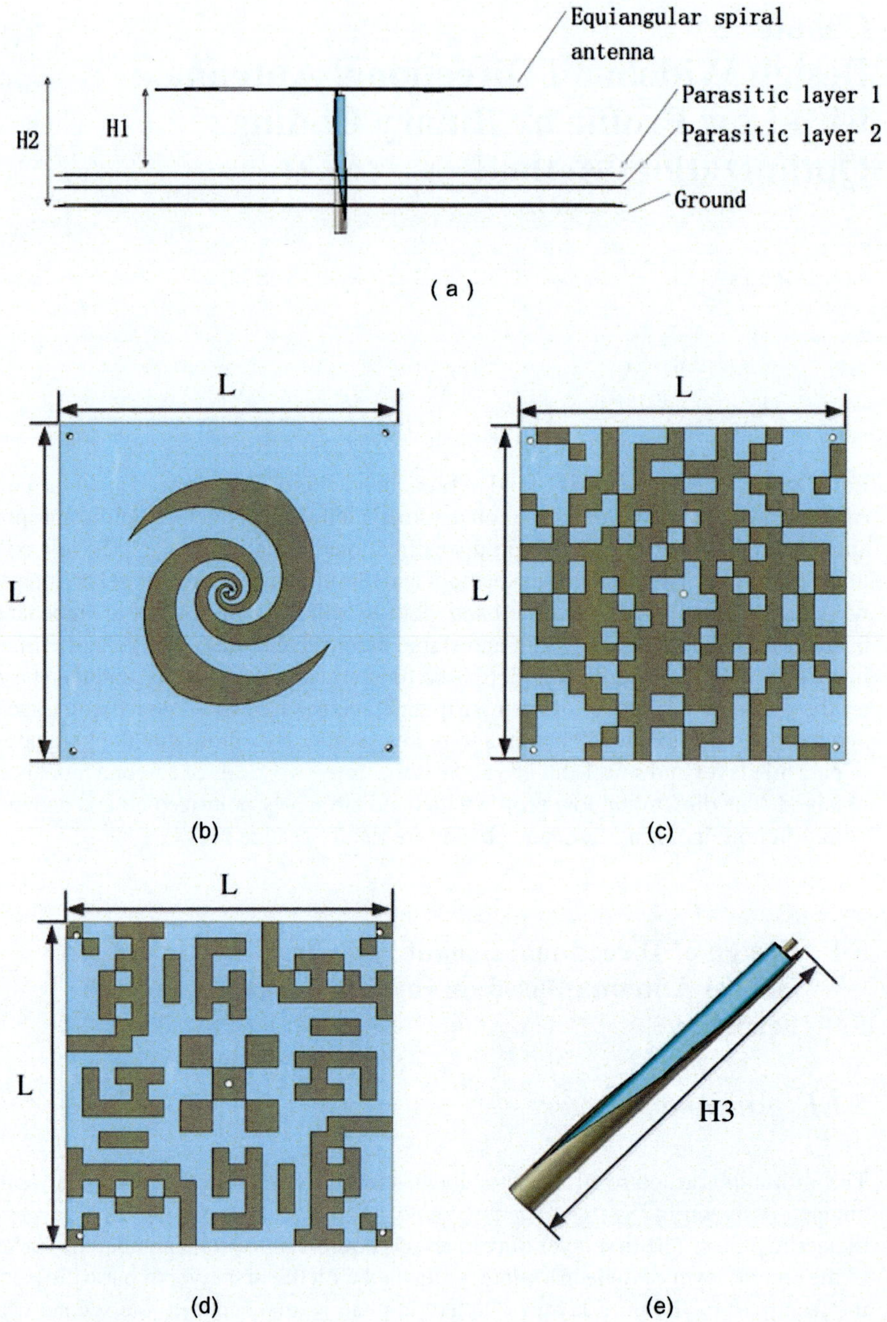

Fig. 5.1 Structure of the antenna **a** Structure of the whole antenna, **b** Equiangular spiral antenna, **c** Parasitic layer1, **d** Parasitic layer2, **e** Cut away style coaxial balun. [3] (reproduced courtesy of The Electromagnetics Academy)

2 mm and 2 mm respectively. After optimization, $a = 0.3$, $r_0 = 1.4$, $\varphi = 3.4\pi$, and the number of turns is 1.7 turns.

The involute coaxial balun [5, 6] is used for the balanced feeding of the two arms of the spiral antenna. For better matching, two fan patches are added at the beginning of the equidistant spiral antenna and they connected with the inner and outer conductors of the involute coaxial cable respectively. At the same time, the thickness of 1 mm of the medium in the output port of the involute coaxial line is cut off, which is the same as the thickness of the first layer of the dielectric plate, so that the involute coaxial balun can contact the two arms of the equidistant spiral antenna after being inserted into the dielectric plate. The optimization of the antenna is mainly to optimize the two parasitic layers [1, 2]. The optimized structures are shown in Fig. 5.1b, c. The whole optimization process uses PSO algorithm to optimize the parameters in CST. The optimization objectives include S11, axial ratio and directivity coefficient, which are multi-objective optimization.

5.1.2 Simulation Results and Analysis

In order to study the effect of two parasitic layers on the antenna performance, we firstly remove the two parasitic layers, leaving only the spiral antenna on the floor, and mark it as Ant1. The distance between the spiral antenna and the floor is fixed. Generally speaking, the bandwidth of pattern stabilization of this kind of patch antenna is relatively limited. In addition, the spiral antenna with two parasitic layers is labeled Ant2.

An ordinary equidistant helical antenna radiates in two directions. When there is a floor, the reflected wave only comes from the reflection on the floor. Traditionally, in order to counteract the reflection of the ground mirror, the distance between the common floor and the spiral antenna is one fourth of the wavelength. After adding two parasitic layers, there are reflected waves in the floor and two parasitic layers. Reasonable structure and layer spacing will help to improve the antenna gain and the stability of the spiral antenna pattern. Two optimized antenna models are simulated by CST Microwave Studio. The specific simulation results are shown in the figure below. Figure 5.2 shows the S parameters of the simulation for Ant1 and ant2. It can be seen that the overall trend of S-parameters is unchanged. In the range of 2.15–12 GHz, $|S_{11}|$ of Ant1 is less than -10 dB, the absolute bandwidth is 9.85 GHz, and the relative bandwidth is up to 5.6:1. In addition, after adding two parasitic layers, Ant2 also maintains broadband characteristics, and parasitic layer has no obvious effect on S parameters.

Figure 5.3 shows the radiation patterns of Ant1 and Ant2 at different frequencies. The results show that the stable directional radiation frequency band of Ant1 is 2–6.8 GHz. The radiation pattern of Ant2 is relatively stable between 2 GHz and 8.7 GHz, and the beam bifurcates between 8.8 GHz and 12 GHz. It can be seen that after adding two parasitic layers, the radiation pattern of Ant2 maintains stable

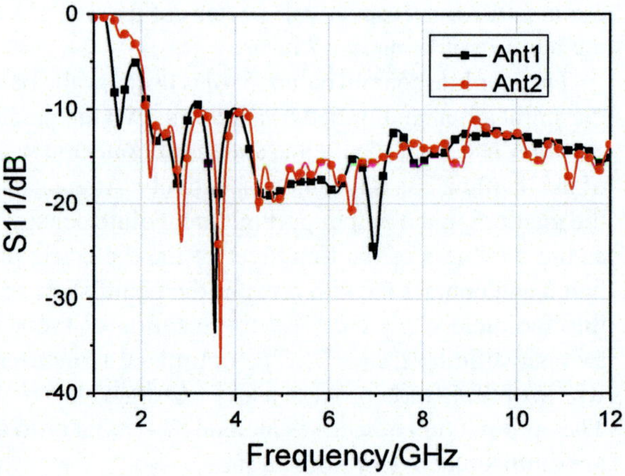

Fig. 5.2 Simulated return loss curves S11 of Ant1 and Ant2. [3] (reproduced courtesy of The Electromagnetics Academy)

directional radiation characteristics between 2 GHz and 6.8 GHz, and it also has very stable directional radiation characteristics between 6.8 GHz and 8.7 GHz.

The simulated main beam direction versus frequency curves of Ant1 and ant2 are shown in Fig. 5.4. It can be clearly seen that in the low frequency band of 2–3.5 GHz, the main beam direction of Ant1 and Ant2 are close to $\theta = 0°$. From 3.5 to 8.5 GHz, The angle range of the main beam direction of Ant1 deviates $\theta = 0°$ is larger than that of Ant2, and the main beam direction of Ant2 changes little in the frequency band.

The 3 dB beamwidths of Ant1 and Ant2 at different frequencies are given in Fig. 5.5, which shows the simulation distribution of the 3 dB beamwidth of Ant1 and Ant2 varies as a function of frequency. In the low frequency range of 2–6 GHz, the 3 dB beamwidths distribution of Ant1 and ant2 are close. In the frequency band of 6–8 GHz, the beam width of Ant2 is wider than that of Ant1.

The simulated curves of axis ratio in the main beam direction of Ant1 and Ant2 changing with frequency are shown in Fig. 5.6. From 4.2 to 8.9 GHz, except around 7.2 GHz point, the axis ratio of Ant1 is almost less than 3 dB, and the corresponding axis ratio bandwidth is 71.2%. From 4.4 to 9.5 GHz, except around 7.2 GHz point, the axis ratio of Ant2 is almost less than 3 dB, and the corresponding axis ratio bandwidth is 73.4%. It can be seen that after inserting two parasitic layers, the frequency band with good axial ratio slightly shifts to higher frequency, and the axial ratio bandwidth increases by 2.4%. This is mainly due to the rotational symmetry of the two parasitic layers and the rotational distribution of the current on the parasitic layer, which improves the circular polarization characteristics of the antenna.

The simulated gain curves of Ant1 and Ant2 versus frequency are presented in Fig. 5.7. From 2 to 4 GHz, the gain of Ant1 and Ant2 are almost the same, but from 4.3 to 7 GHz, the gain of Ant2 is higher than that of Ant1. From 7 to 12 GHz, the gain of Ant1 is relatively high, but the radiation direction of the main beam of Ant1 is unstable and begins to twist.

Fig. 5.3 Simulated far field patterns **a** Ant1, **b** Ant2. [3] (reproduced courtesy of The Electromagnetics Academy)

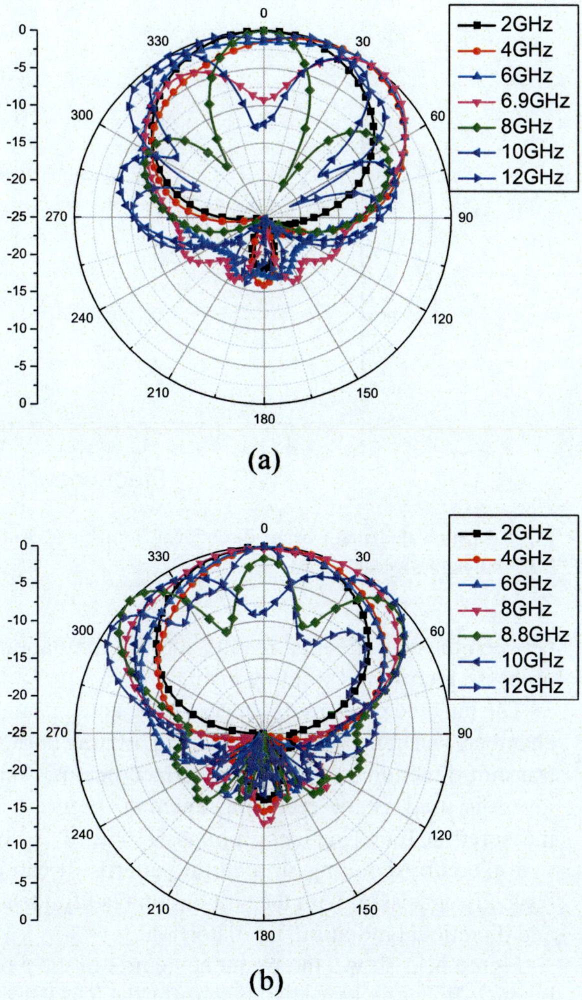

5.1.3 Analysis of Simulation and Measured Results

The optimized prototype of the Ant2 is manufactured as shown in Fig. 5.8.

The antenna was measured by Agilent E8361C vector network analyzer, and the variation curve of the S parameters of the manufactured antenna with frequency is given in Fig. 5.9, which is compared with the simulation results. In Fig. 5.9, the frequency range of S11 less than −10 dB of the experiment antenna is 2.05–12 GHz; The measured and simulated results show that the S parameter is less than −10 dB in the frequency range of 2.15–12 GHz. In the lower frequency range of 2.05–2.15 GHz, the experiment results are better than the simulation results, but the overall trend of measurement and simulation is basically close. The difference

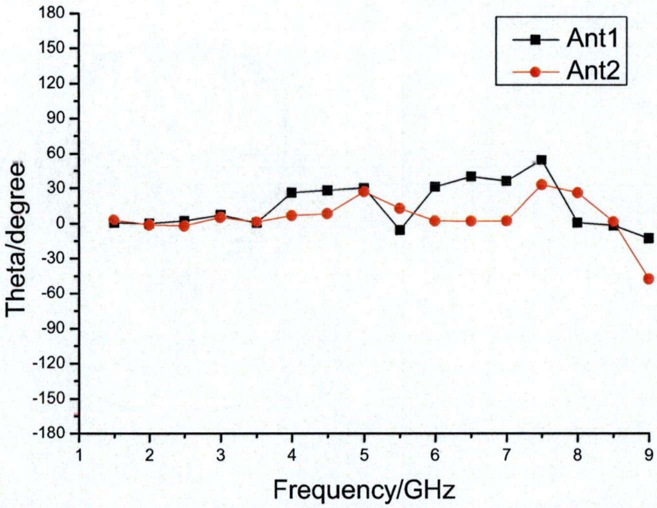

Fig. 5.4 Simulated main beam direction of Ant1 and Ant2. [3] (reproduced courtesy of The Electromagnetics Academy)

between the experimental results and the simulation results mainly comes from the processing error and welding error.

The far-field patterns of the prototype antenna are tested in a microwave anechoic chamber. The transmitting antenna is a rotating horn antenna, which is mainly used to transmit directional circularly polarized beams. Ant2 with two optimized parasitic layers is used as the receiving antenna. Figure 5.10 shows the far-field patterns measured at four frequency points (f = 2, 4, 5 and 7 GHz), and compare them with the corresponding simulation patterns. It can be seen that the test results are basically consistent with the simulation results, the antenna far-field pattern is stable and directional radiation, no distortion.

Figure 5.11 shows the frequency curve of the axis ratio measured and simulated by ant2. It can be seen that except for the 7.2GHhz frequency point, the frequency range of measured axial ratio less than 3.8 dB is 4.6–8 GHz. The experimental results are in keeping with the simulation results.

Figure 5.12 shows the curves of the measured and simulated gain versus frequency. The results show that the simulation results are basically consistent with the measured. The measured gain range is 6.22–9.5 dBi. In the high frequency range of 6.2–8 GHz, the measured gain is 0.2–1 dB higher than the simulated. This is mainly because in the actual test, there is a certain height difference between the receiving antenna and the transmitting horn. Certainly, the error of the chamber test system also has a certain influence on the test results (±0.5 dB).

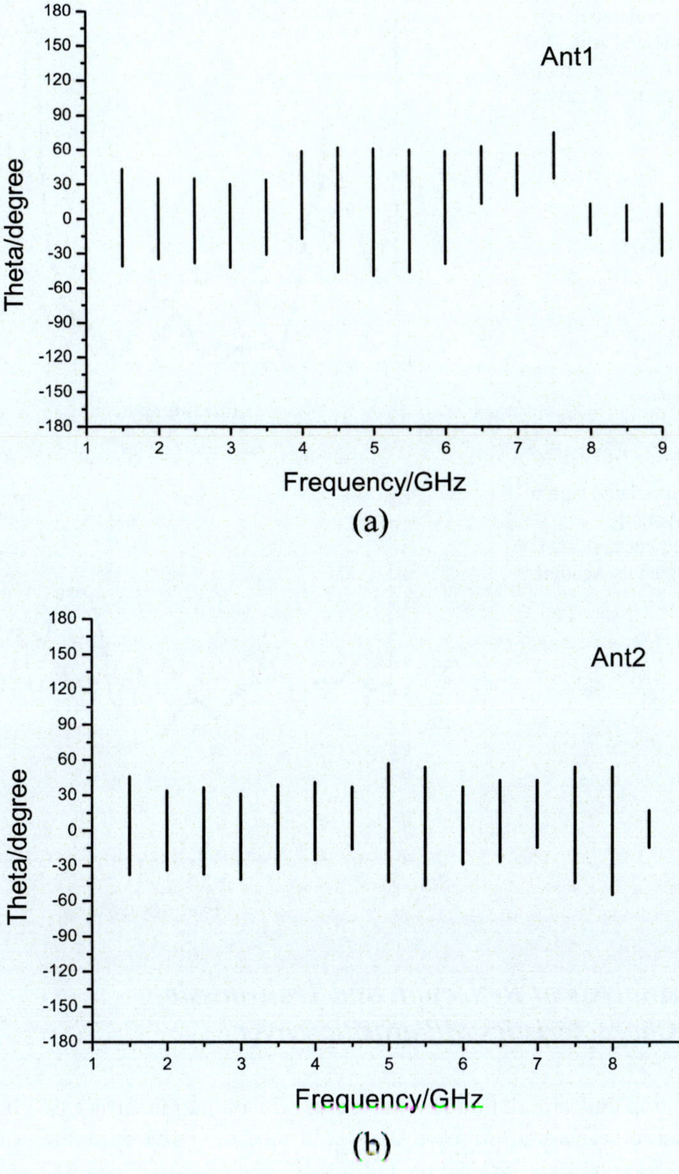

Fig. 5.5 Simulated 3 dB beam width distribution of **a** Ant1 and **b** Ant2. [3] (reproduced courtesy of The Electromagnetics Academy)

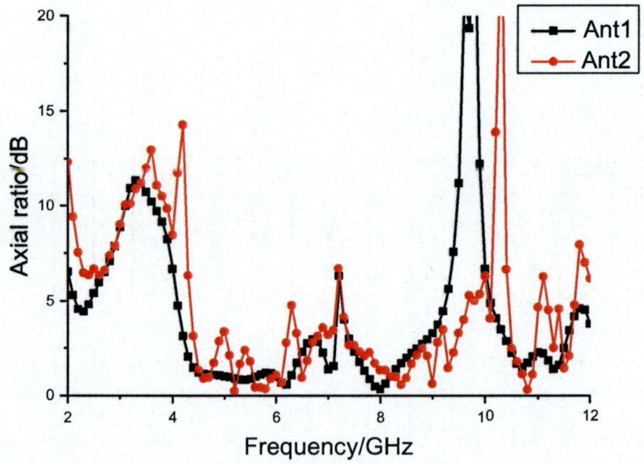

Fig. 5.6 Simulated axial ratio of Ant1 and Ant2. [3] (reproduced courtesy of The Electromagnetics Academy)

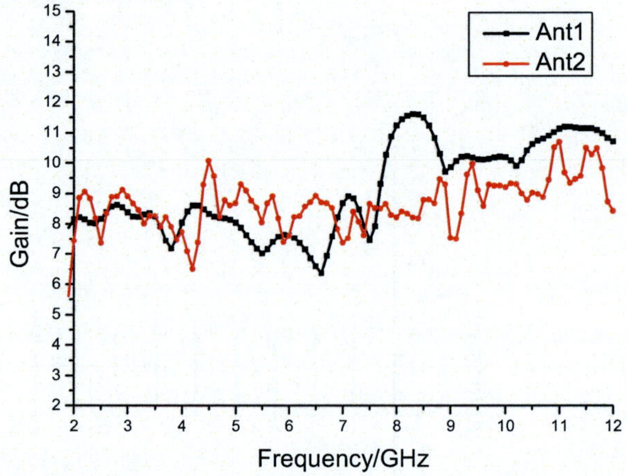

Fig. 5.7 Simulated gain of Ant1 and Ant2. [3] (reproduced courtesy of The Electromagnetics Academy)

5.1.4 Analysis of Reflection and Transmission Characteristics of Parasitic Layer

From the previous simulation results, it can be seen that inserting two binary coded grid parasitic layers can improve the stability of the pattern, and slightly improve the axial ratio and gain. So, we try to illustrate the working principle of the antenna prototype through the simulation of different structures. Specifically, for the three cases of only limited ground, only two optimized binary coded grid parasitic layer and the combination case of the two binary coded grid parasitic layers and limited floor, circularly polarized plane wave is used for vertical irradiation, and the power density distribution of reflected wave and transmitted wave in these three cases are simulated and calculated. The results are shown in Figs. 5.13, 5.14 and 5.15. The size and spacing of the floor and parasitic layers in the simulation are exactly the

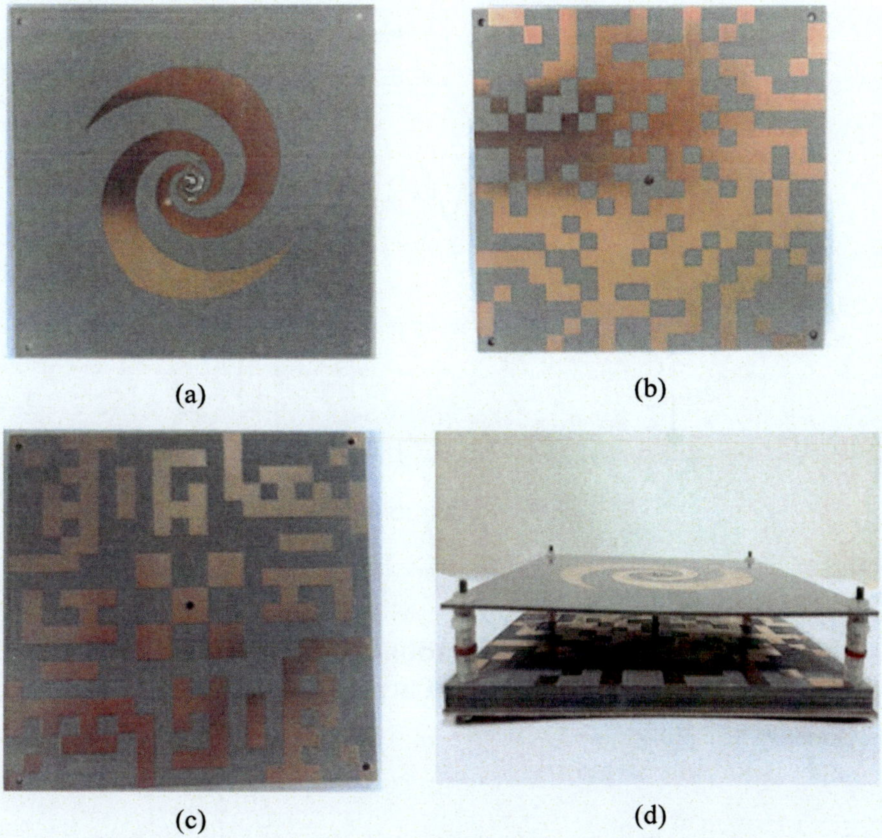

Fig. 5.8 Photograph of the fabricated antenna **a** Equiangular spiral antenna, **b** Parasitic layer 1, **c** Parasitic layer 2, **d** Side view. [3] (reproduced courtesy of The Electromagnetics Academy)

same as those in the optimized antenna prototype structure. Obviously, in Fig. 5.13, there is only a limited propagation time, and the power density of reflected wave is larger at some places in the edge and middle region, while the power density of transmitted wave is larger at the edge. In Figs. 5.14 and 5.15, the power density of the reflected wave is larger in the center, and has the characteristics of rotational symmetry on the whole. The power density of the transmitted wave is relatively large at the edge. Through the comparison of these cases, it can be seen that adding the two optimized binary coded grid parasitic layers can focus the intermediate energy, improve the stability of the graphics, and increase the gain. The rotationally symmetric distribution of power density also helps to improve the axial ratio.

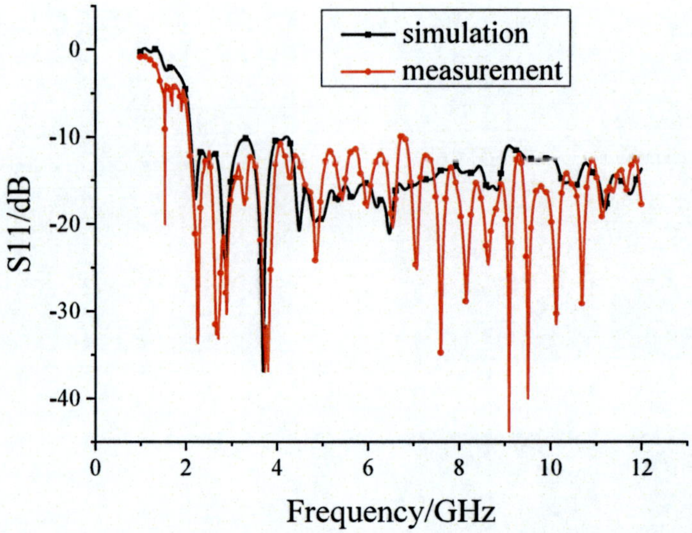

Fig. 5.9 Simulated and measured S11 of the fabricated Ant2

5.2 Design of Directional Broadband Circularly Polarized Helical Antenna Based on Parallel Two-Line Feed

5.2.1 Antenna Structure

The structure of the broadband circularly polarized equiangular spiral antenna based on double parallel line feed is shown in Fig. 5.16. Similar to the previous antenna, the antenna has four layers of metal and three layers of dielectric plate. The first metal layer is an equidistant spiral antenna, the first parasitic patch is attached to the second dielectric plate, the second parasitic patch is attached to the third dielectric plate, and the floor is attached to the back of the third dielectric plate. The dielectric constant of the three-layer dielectric plate is the same ($\varepsilon_{r1} = \varepsilon_{r2} = \varepsilon_{r3} = 2.65$), and the thickness of the dielectric plate is 1 mm, 2 mm and 2 mm respectively. To further adjust the reflected wave phase to strengthen the upward radiation, there is vacuum between the first dielectric layer and the firt parasitic layer. The line parameters of the optimized isometric spiral curve are $a = 0.29$,$r_0 = 0.82$ mm, $\varphi = 3.1\pi$, that is, the number of turns is 1.55. The Balun made of parallel double microstrips is fed through a microstrip line and connected to the starting point of the two arms in the center of the equiangular spiral antenna. Balun's parameters are $W1 = 0.8$ mm, $W2 = 1.8$ mm, $W3 = 7.8$ mm, $H = 16$ mm. There are round holes with radius of 4.5 mm in the center of parasitic layer 1 and parasitic layer 2 and the, so that the Balun made of parallel double microstrip can pass through.

To verify the performances of the optimized directional broadband circularly polarized helical antenna based on parallel two-line feed, we construct a comparing

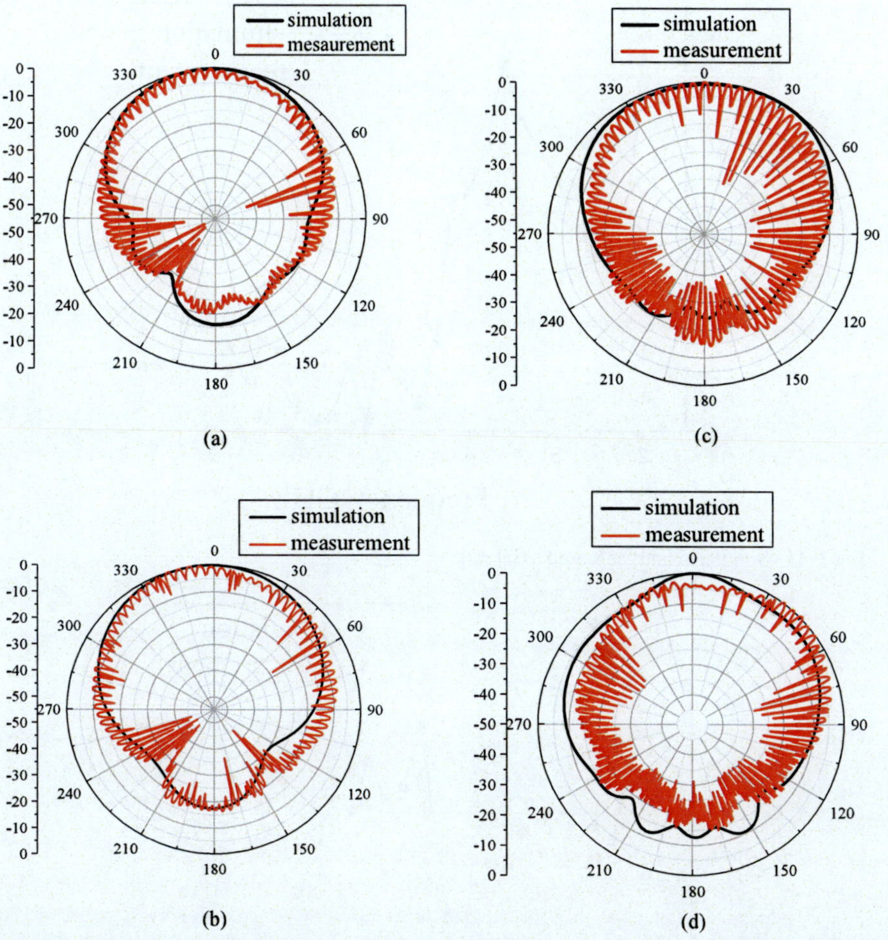

Fig. 5.10 Simulated and measured far field pattern of the fabricated Ant2 **a** 2 GHz, **b** 3 GHz, **c** 5 GHz, **d** 7 GHz

object, which just include the equiangular spiral antenna and ground, called as Ant3, and called the optimized directional broadband circularly polarized helical antenna based on parallel two-line feed as Ant4.

5.2.2 Simulation Results and Analysis

After simulation and optimization in CST Microwave Studio, the simulated return loss S11 of Ant3 and Ant4 changes with frequency is shown in Fig. 5.17. The impedance bandwidth of Ant1 with S11being less than −10 dB is from 3 to 9 GHz;

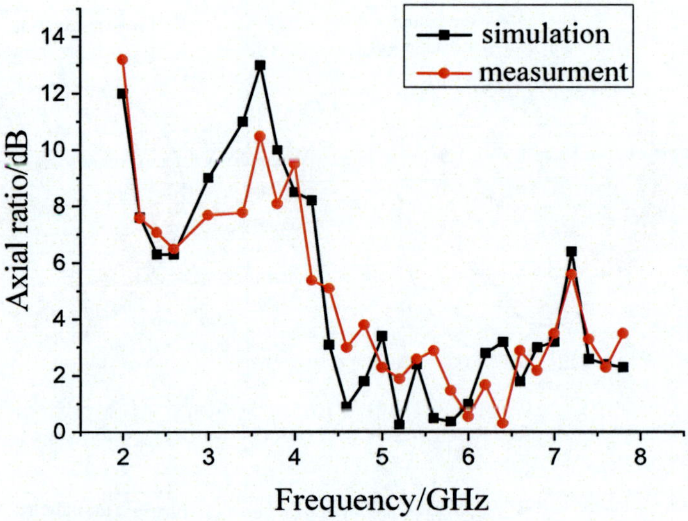

Fig. 5.11 Simulated and measured axial ratio of Ant2

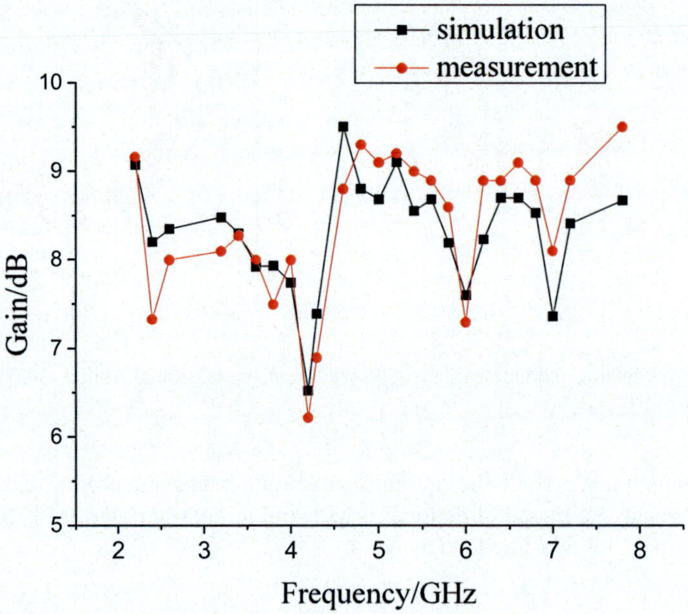

Fig. 5.12 Simulated and measured gain of Ant2

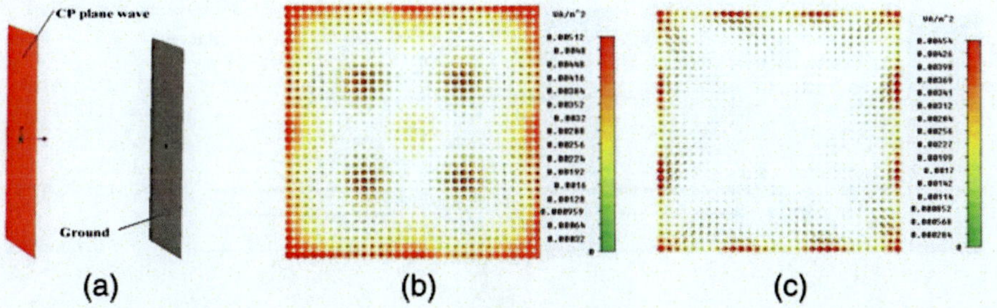

Fig. 5.13 **a** Circularly polarized plane wave radiates the limited ground, **b** Simulated reflected wave power flow density at 7.5 GHz, **c** Simulated transmitted wave power flow density at 7.5 GHz

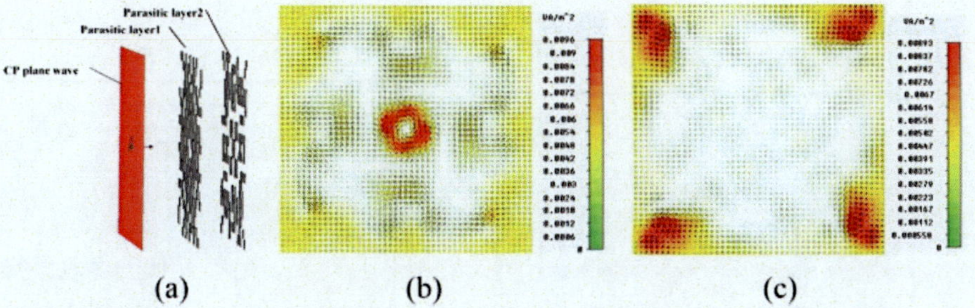

Fig. 5.14 **a** Circularly polarized plane wave radiates the parasitic layers, **b** Simulated reflected wave power flow density at 7.5 GHz, **c** Simulated transmitted wave power flow density at 7.5 GHz

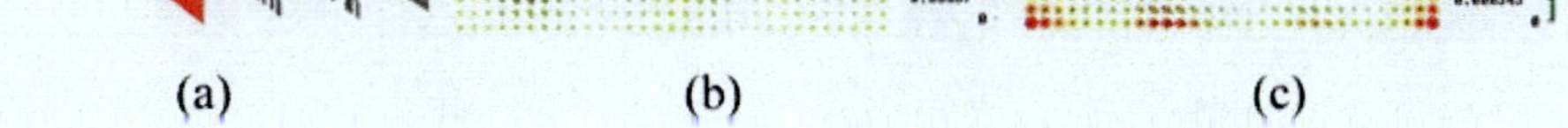

Fig. 5.15 **a** Circularly polarized plane wave radiates the parasitic layers with limited ground, **b** Simulated reflected wave power flow density at 7.5 GHz, **c** Simulated transmitted wave power flow density at 7.5 GHz

but it is in the 4–9 GHz band to Ant4. It means that inserting two parasitic layers deteriorate the return loss S11 of Ant4 at a low frequency band of 3 GHz to 4 GHz.

The simulated far field patterns of Ant3 and Ant4 at 4, 5, 6, 7 and 8 GHz are given in Fig. 5.18. In Fig. 5.18a, the main beam of Ant3 at 7 GHz deviated from the direction of $\theta = 0°$, the normal direction of Ant3, and the far field pattern has

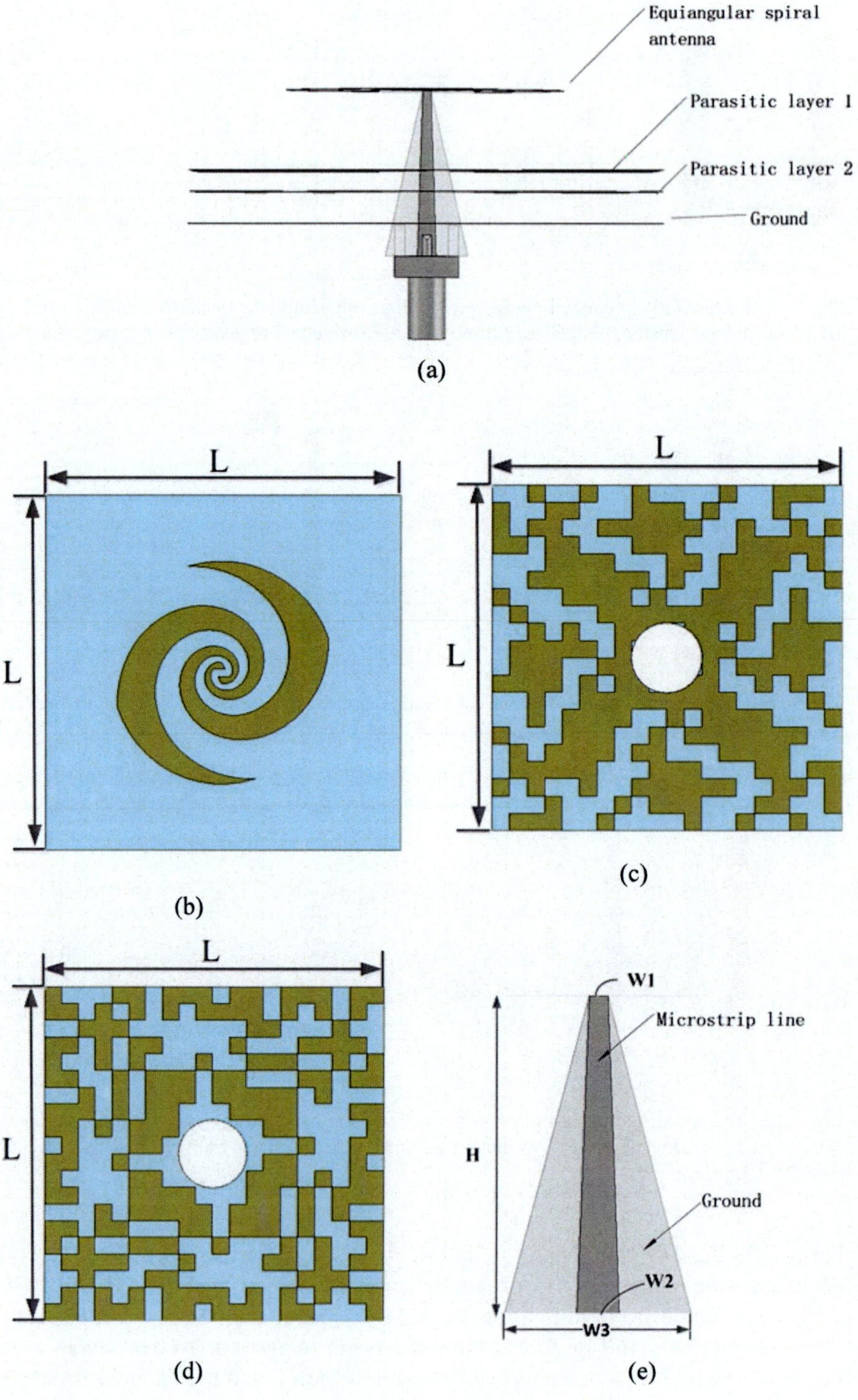
Equiangular spiral antenna
Parasitic layer 1
Parasitic layer 2
Ground
(a)
L
L
(b)
L
L
(c)
L
L
(d)
W1
Microstrip line
H
Ground
W2
W3
(e)

◂**Fig. 5.16** Structure of the antenna **a** Structure of the whole antenna, **b** Equiangular spiral antenna, **c** Parasitic layer1, **d** Parasitic layer2, **e** balun. Figure reproduced with permission from Ref. [4], © 2017 IJAP

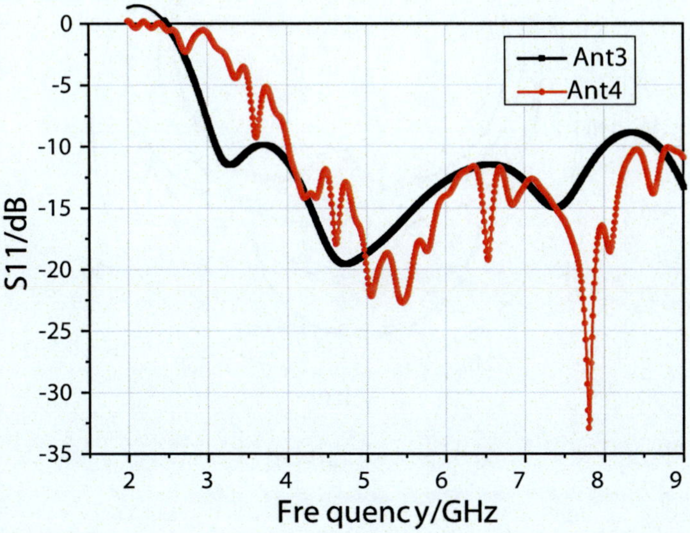

Fig. 5.17 Simulated S11 of Ant3 and Ant4. Figure reproduced with permission from Ref. [4], © 2017 IJAP

obvious bifurcation at 8 GHz around the direction of θ = 0°. In Fig. 5.18b, all the main beams point to the direction of θ = 0°, the normal direction of Ant4. It is obvious that the directional radiation performance of Ant4 is better than Ant3 over the band of 4–8 GHz, which means that the inserted two parasitic binary coded mesh layers improves the stability and directional radiation performance of Ant4.

The measured axial ritio cuve and simulated axial ritio cureve over frequency are given in Fig. 5.19. It can be seen that the circular polarization band of Ant3 with axis ratio being less than 3 dB is from 4 to 4.6 GHz, and it is from 4.3 to 5.2 GHz to the Ant2. So, we can deduce that the inserted two parasitic binary coded mesh layers widen the CP bandwidth from 13.9% of Ant3 to 18.9% of Ant4, and the CP band of Ant2 shifts to the high frequency with respect to axial ritio of the Ant3 and Ant4.

The simulated gain versus frequency cuves of Ant3 and Ant4 are presented in Fig. 5.20. It is clear that the gain of Ant1 is 5.6–7.35 dB in the frequency band of 4–8 GHz, while the gain Ant2 is 6.3–9.37 dB, which is higher than the overall gain of Ant1. The gain improvement is just from the parasitic binary coded mesh layers.

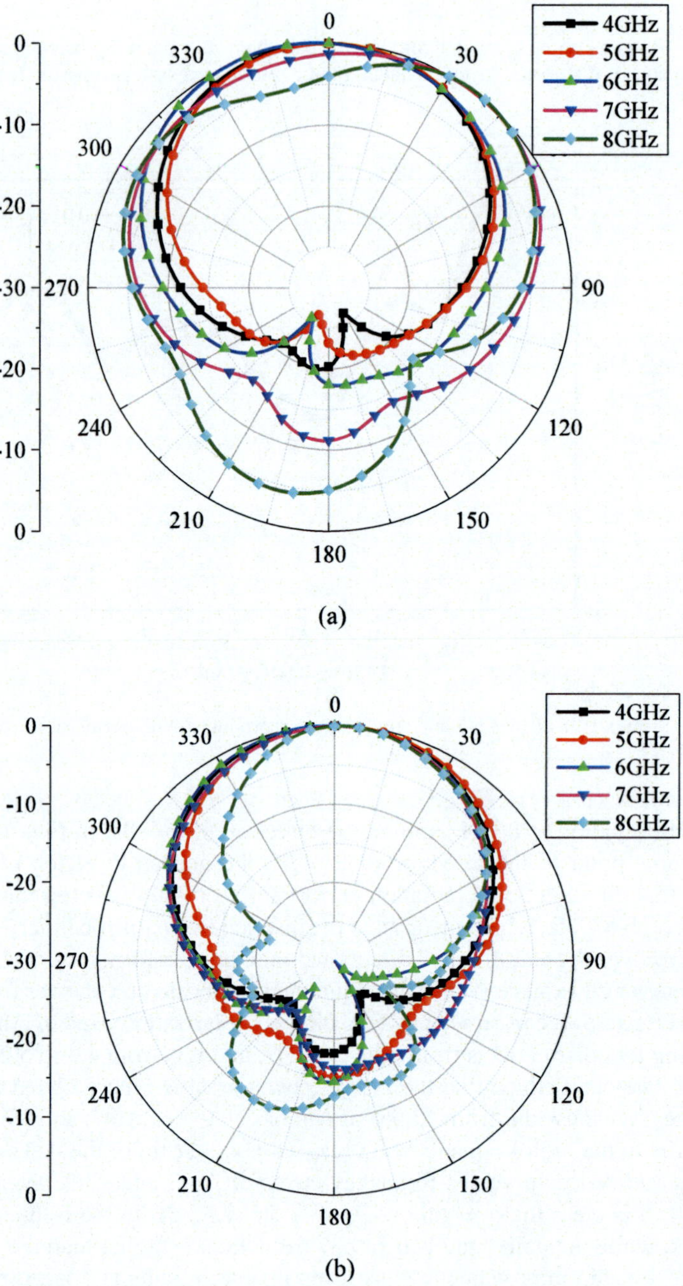

Fig. 5.18 Simulated far field patterns **a** Ant3, **b** Ant4. Figure reproduced with permission from Ref. [4],© 2017 IJAP

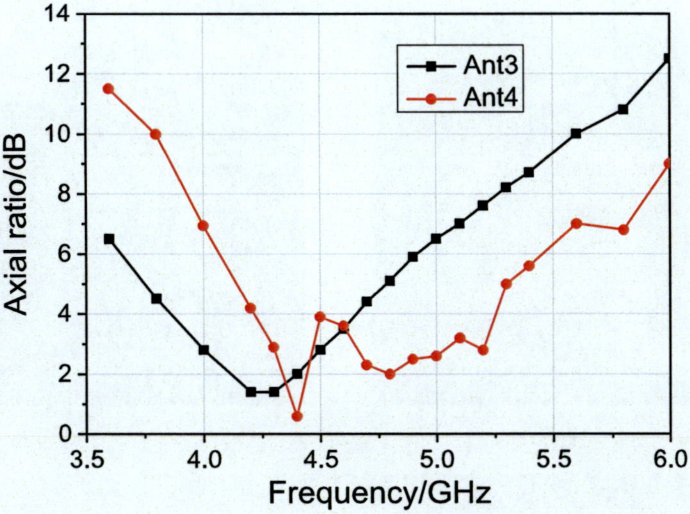

Fig. 5.19 Simulated axial ratio of Ant3 and Ant4

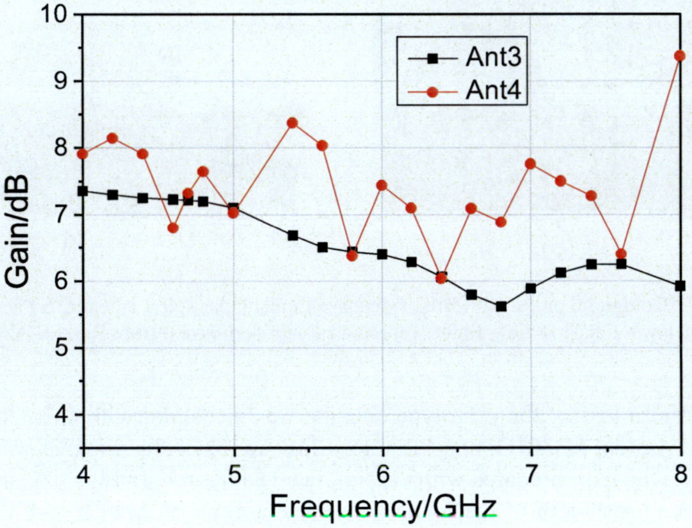

Fig. 5.20 Simulated gain of Ant3 and Ant4

5.2.3 The Experiment Results Analysis

The real prototype of Ant4 is manufactured according to the optimized antenna model structure, which is given in Fig. 5.21.

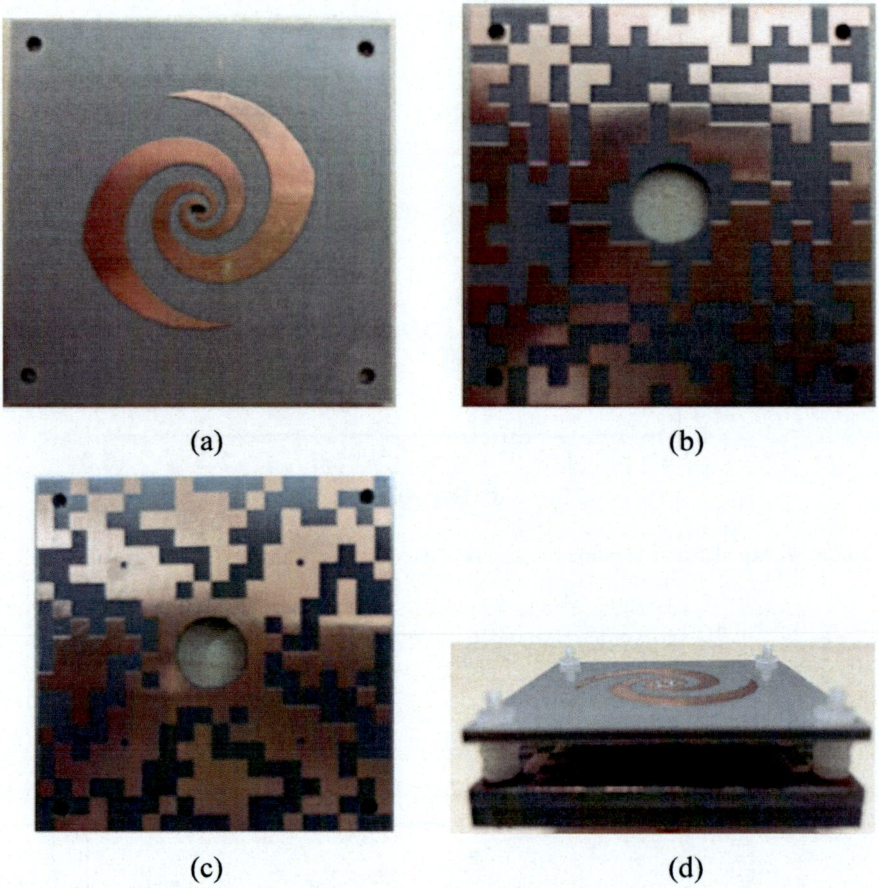

Fig. 5.21 Photograph of the fabricated antenna **a** Equiangular spiral antenna, **b** Parasitic layer 1, **c** Parasitic layer 2, **d** Side view. Figure reproduced with permission from Ref. [4], © 2017 IJAP

The return loss of the prototype antenna was measured with an vector network analyzer Agilent E8361C, and the return loss curve of the antenna was shown in Fig. 5.22, which is compared with the simulated S11 curve. In Fig. 5.22, the available impedance bandwidth of the manufactured prototype of Ant4 is 4–8 GHz with S parameter being less than −10 dB, which is consistent with the trend of the simulation result.

The far-fields of the manufactured prototype of Ant4 were measured in a microwave chamber. The rotating horn antenna is set as the transmitting antenna, and the manufactured prototype of Ant4 was used as the receiving antenna. Figure 5.23 shows the measured normalized patterns of the prototype antenna at $f = 4.2$, 4.6, 5. GHz, 5.8 GHz, and the simulated patterns are compared in the figure toos. It is clear that the measured far fields results are in agreement with the simulation results as a whole, and the directional beams are upward and stable without distortion.

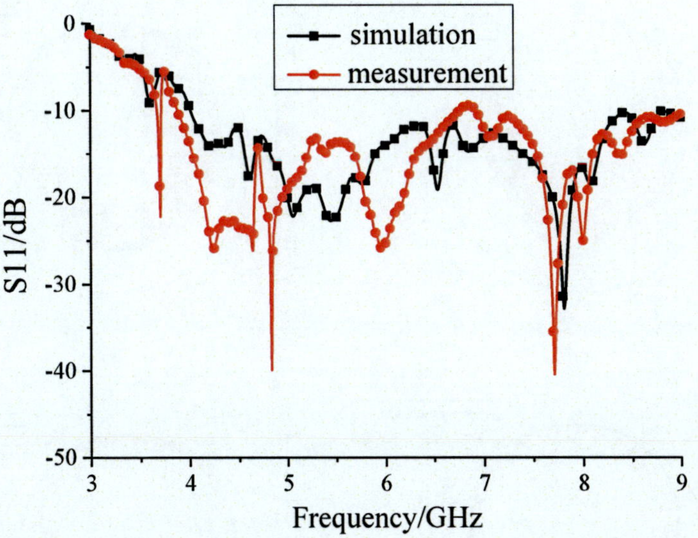

Fig. 5.22 Simulated and measured S11 of the fabricated Ant4

The measured and simulated axial ratio versus frequency curves of the prototype antenna are shown in Fig. 5.24. The measured axial ratio is mostly less than 3 dB in the band of 4.5–5.4 GHz. Although the measured and the simulated axial ratio curves are different, the overall trend of the curves are close, and the difference between these two axial ratio curves is mainly due to processing errors and experiment errors.

The measured and simulated gain versus frequency curves of the prototype antenna are given in Fig. 5.25. In the stable directional radiation band of 4–8 GHz, the measured gain range is 6.1–8.3 dB, which is close to the simulation result, and the trends are basically same.

5.2.4 Analysis of Reflection and Transmission Characteristics of Parasitic Layer

As mentioned above, the introduction of two binary coded parasitic grid layers can facilitate optimization and improve or increase the stable directional pattern bandwidth, axial ratio bandwidth and gain of the isometric spiral antenna. We can explain the reason of the performance improvement with the different structure radiated by circularly polarized plane wave. Specifically, a circularly polarized plane wave is used as a source to irradiate a limited ground, a combination of two parasitic layers or a combination of two parasitic layers and a limited floor. Figures 5.26, 5.27 and 5.28 show the simulated power flow density distributions of reflected wave and transmitted wave. In the simulation model, the size of each layer, the limited floor and the

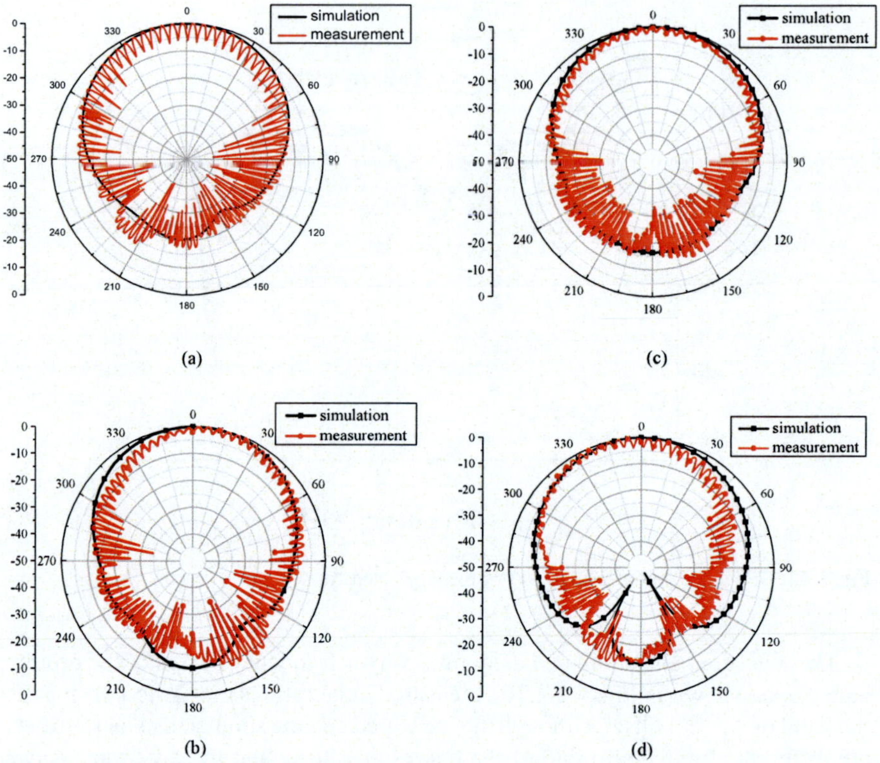

Fig. 5.23 Simulated and measured far field pattern of the fabricated Ant4 **a** 4.2 GHz, **b** 4.6 GHz, **c** 5.1 GHz, **d** 5.8 Ghz

distance between them are exactly the same as those in the optimized antenna structure of Ant4, and there is a circular through hole in the middle of the limited ground, which is exactly the same as the optimized antenna structure of Ant4. In Fig. 5.26, in a limited calculating time, the reflected wave power flow density is relatively large in the edge and middle region. In Figs. 5.27 and 5.28, the point with higher power flow density of the reflected wave appears in the center, and the distribution of power flow density of the reflected wave is rotationally symmetric. The power density of forward propagation wave is relatively large at the edge. It can be seen that adding two binary coded parasitic grid layers can focus the energy in the center, which is helpful to improve the stability and gain of the pattern. The rotational symmetry of the stable power flow density distribution also increases the axial bandwidth of the antenna.

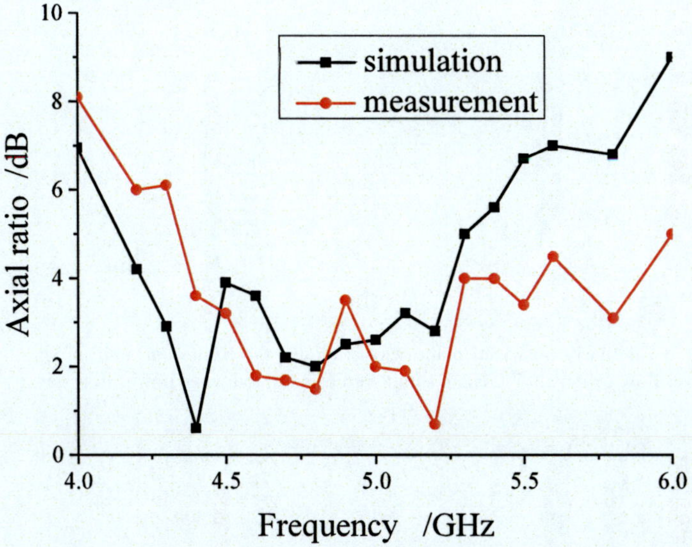

Fig. 5.24 Simulated and measured axial ratio of Ant4

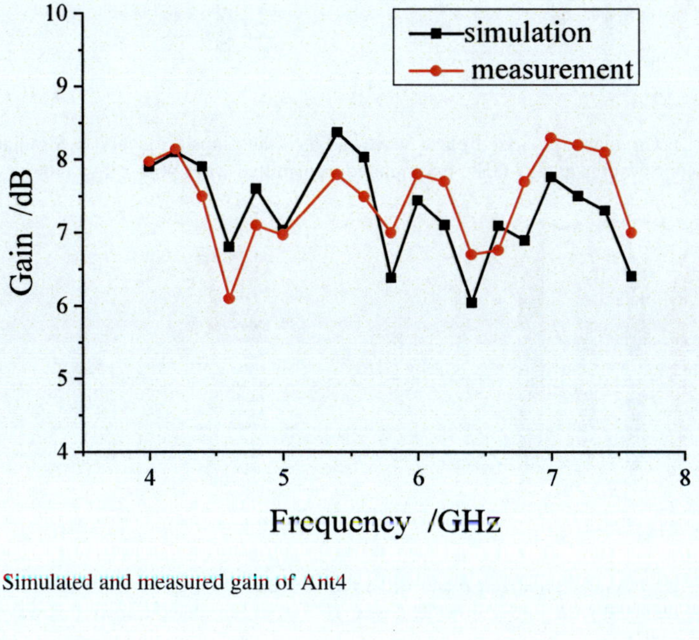

Fig. 5.25 Simulated and measured gain of Ant4

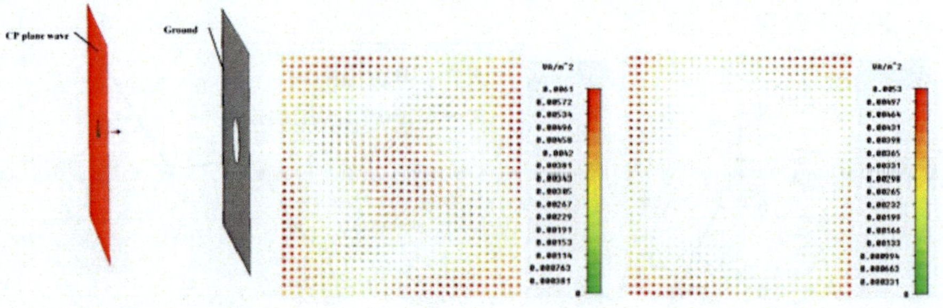

(a) (b) (c)

Fig. 5.26 **a** Circularly polarized plane wave radiates the limited ground, **b** Simulated reflected wave power flow density at 7 GHz, **c** Simulated transmitted wave power flow density at 7 GHz

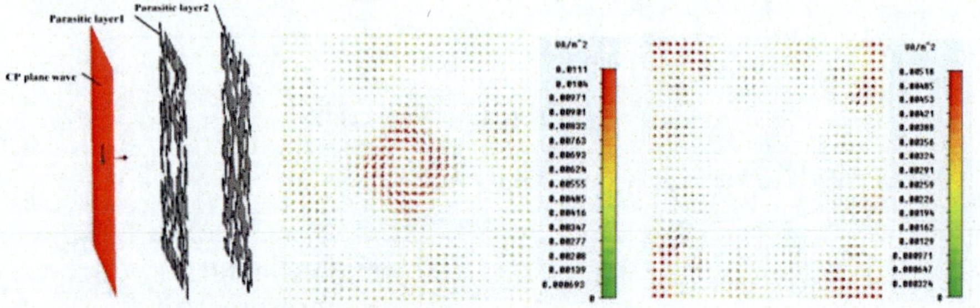

(a) (b) (c)

Fig. 5.27 **a** Circularly polarized plane wave radiates the parasitic layers, **b** Simulated reflected wave power flow density at 7 GHz, **c** Simulated transmitted wave power flow density at 7 GHz

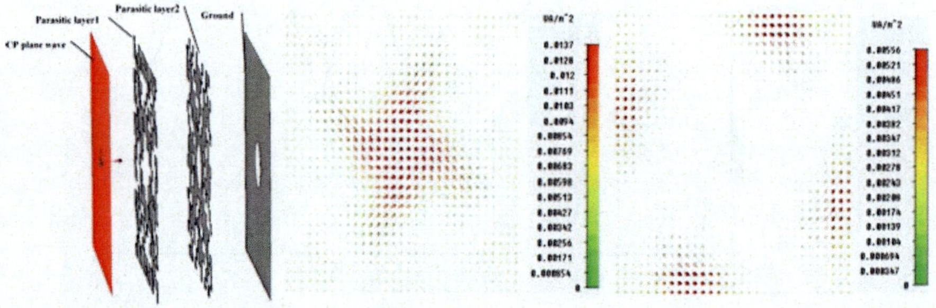

(a) (b) (c)

Fig. 5.28 **a** Circularly polarized plane wave radiates the parasitic layers with limited ground, **b** Simulated reflected wave power flow density at 7 GHz, **c**) Simulated transmitted wave power flow density at 7 GHz

5.3 Summary

This chapter firstly introduces the basic knowledge of broadband balun, and modifies the rotational symmetry of microstrip antenna optimization model. Based on different feeding modes, two directional circularly polarized UWB antennas with different operating frequency bands are optimized by introducing binary coded grid parasitic layer and equiangular spiral antenna. The simulation results are compared with the equidistant spiral antenna without parasitic layer, and the principle of their broadband directional circularly polarized radiation is analyzed. The measured and simulated results of the two antenna prototypes are in good agreement respectively, which also proves the accuracy of the design.

The first equiangular spiral antenna is fed by involute coaxial cable. The measured impedance bandwidth ($|S_{11}| < -10$ dB) is 5.8:1 (2.06–12 GHz), and the bandwidth with stable directional radiation pattern is 2–8.7 GHz. The frequency range of axial ratio less than 3.8 dB is from 4.6 to 8 GHz. The measured gain is between 6.22 and 9.5 dB. The profile of the whole antenna is 19.3 mm, which is about one eighth of the lowest operating frequency. Through the simulation of the reflection and transmission characteristics of the parasitic layer and the floor, it is found that the distribution of power flow density is consistent with the design expectation of broadband directional circularly polarized radiation.

The measured impedance bandwidth ($|S_{11}| < -10$ dB) is 2:1 (4–8 GHz). The bandwidth with stable directional radiation pattern is 4–8 GHz. The measured axial ratio is less than 3 dB in the range of 4.5–5.3 GHz. The test gain is 6.1–8.3 dB. The profile of the whole antenna is 17 mm, which is about a quarter of its lowest operating frequency. Then the reflection and transmission characteristics of the parasitic layer and the floor are simulated and compared. It is found that the distribution of power flow density is consistent with the design expectation of broadband directional circularly polarized radiation.

References

1. H. Wu, W. Liu, J. Chen, S. Liu, in *A New Particle Swarm Optimization and Its Application to Antenna Design*. Outstanding Academic Papers by Students (2009)
2. H. Wu, J. Geng, R. Jin, J. Qiu, W. Liu, J. Chen, S. Liu, An improved comprehensive learning particle swarm optimization and its application to the semi-automatic design of antennas. IEEE Trans. Antennas Propag. **57**(10), 3018–3028 (2009)
3. X. Liu, J. Geng, X. Liang, R. Jin, C. Zhang, K. Wang, An improvement to directional equiangular spiral antenna with wide CP band. High Gain Low Profile Progr. Electromagn. Res. C **48**, 53–60 (2014)
4. X. Zhao, J. Geng, R. Jin, Y. Jin, X. Liu, W. Yin, Topological design of planar circularly polarized directional antenna with low profile using particle swarm optimization. Int. J. Antenna Propag. **2017**(Article ID 4983724), 12 pages (2017)

5. J. Lao, R.H. Jin, J.P. Geng, UWB spiral antenna with parabolic reflector, XXIX URSI General Assembly, Chicago, Illinois, USA (2008)
6. J.W. Duncan, V. P. Minerva, 100 : 1 bandwidth balun transformer *Proceedings of the IRE*, Vol. 48, No. 2, 156–164 (1960)

Chapter 6
Design mm Wave Antenna by Binary Coding Optimization Method

6.1 Introduction

With the development of society, people's demand for high-speed wireless transmission is increasing, and wireless communication has achieved rapid development, but the spectrum resources are becoming increasingly scarce, especially the spectrum resources below 6 GHz are scarce, but the demand for higher transmission rate is also strong in today's society. Therefore, the development of 5G/6G antenna in millimeter wave band becomes the key to solve these contradictions. As 5G technology will become the mainstream of wireless communication, the expected maximum transmission rate can reach 10Gbps. The absolute bandwidth of millimeter wave band is sufficient, and it can carry more information capacity when working in millimeter wave band. In fact, with the development of wireless communication technology from the first generation of analog communication to today's 5G, the transmission rate has been increased from the initial hundreds of bps to 100Mbps of 4G, which can support multimedia services, digital broadcasting, multi video conference, etc., making the development of today's society into the fast lane. Therefore, 5G supporting 10Gbps transmission rate is the inevitable trend of wireless communication development.

Millimeter wave communication has become the key of emerging 5G mobile communication. Millimeter wave antenna has low profile and large absolute bandwidth. Millimeter wave broadband antenna has compact structure and is easy to be conformal. It can easily support gigabit per second (Gbps) data rate in wireless channel, greatly improving channel capacity and anti-interference ability of antenna; The circular polarization characteristics of millimeter wave antenna can reduce the multipath effect and improve the occurrence of polarization mismatch in complex environment; Using millimeter wave antenna elements to form an array can effectively improve the gain of the whole antenna system and reduce the path loss. These characteristics of millimeter wave antenna can ensure the normal operation of 5G system electromagnetic information, and it is an ideal choice to support high data rate in 5G wireless communication [1–3].

J. Geng and R. Jin, *Antenna Optimization and Design Based on Binary Coding*, Modern Antenna, https://doi.org/10.1007/978-981-16-7965-0_6

The innovation of the new generation of wireless communication system needs the global cooperation. Many people in the world have been studying 5G and millimeter wave communication, and given a lot of measurement methods, lessons and research results [4, 5]. The path loss and rain attenuation of millimeter wave are large. In the heavy rain environment (25 mm/HR), there will be 7 dB/km of electromagnetic wave propagation at 28 GHz. In the urban environment, if a relay station is set every 200 m, the electromagnetic energy loss will be reduced to 1.4 dB [6]. The energy loss of 28 GHz signal is only 0.018 dB per 200 m in the atmosphere [7]. In other words, in short-range communication, the atmospheric loss and rain loss are very small for millimeter wave.

R. B. Waterhouse et al. designed a stacked antenna with aperture coupling feed in 2003. The antenna is fed by aperture coupling feed. By adding parasitic patches, the antenna covers the whole Ka band (26–40 GHz), with a relative bandwidth of 42.2%. The array gain based on millimeter wave unit reaches 14–17 dB, which has good application value [8]. In 2018, Mohsen khalily et al. used microstrip direct feed, and reduced coupling through inverse rotating feed, and proposed a 16 element array of booth. It can work at 24.35–31.13 GHz, with relative bandwidth of 24.4%, array gain of 19.88dbi and maximum scanning angle of 50°. Jonas kornprobst et al. designed a millimeter wave antenna with broadband and wide angle scanning characteristics in 2017. The antenna is fed by coupling and excites two similar resonant modes. The bandwidth of the antenna is 34.1–38.9 GHz, and the relative bandwidth is 13.1%. In addition, two parasitic patches are placed on both sides of the patch, which can increase the gain of bias to the horizontal plane and achieve the performance of wide beam. The half power beam width (HPBW) of the antenna is 100°–125° [9].

Chun Xu Mao et al. designed a third-order vertical coupling antenna, which stacked the microstrip feed structure, U-shaped slot and rectangular patch from bottom to top to realize the simultaneous resonance of the third-order resonant mode in 2017. The antenna works at 22–32 GHz and the relative bandwidth reaches 37%. The gain of the array can reach more than 20 dB [10]. Karthikeya gulur sadananda et al. designed a dual polarized millimeter wave antenna unit in 2019. The mutual coupling between the two ports is only—22 dB at 28ghz. The gain of the antenna is as high as 9.2–9.6 dbi, with good dual polarization and high gain characteristics [11].

In reference [12], the U-slot structure is used to broaden the bandwidth, and the relative bandwidth of the antenna in millimeter wave band is 21.92%. In [13], E-shaped patch is used to design millimeter wave broadband antenna, which can cover 31.6–40 GHz, 23.5% of the relative broadband. A dual polarized magnetoelectric dipole antenna with a bandwidth of 65.9% is presented in Ref. [14]. It is excited by T-shaped strips with two input ports, but it has a complex three-dimensional structure.

It can be seen that under the promotion of 5G, millimeter wave antenna has become a research hotspot. The existing millimeter wave antenna has realized the characteristics of high gain, broadband, dual polarization and high gain array. However, the performance of most millimeter wave antennas is relatively single. It is a difficult and attractive problem that how to integrate these high-performance antennas into one antenna to achieve broadband wide-angle dual polarization, efficient array formation and large angle scanning in millimeter wave band. In this chapter, a broadband millimeter wave antenna with wide half power radiation width is designed by using binary coded grid optimization technology, and the unit is expanded into an array to achieve large angle scanning.

6.2 Binary Coding Proccess

According to the method of binary coding of meshes in Chapt. 2, the antenna is coded and optimized. The basic binary coding model of the millimeter wave antenna is shown in Fig. 6.1. The antenna model generally includes seven layers: discrete air layer 1, patch layer 1, dielectric layer 1, air layer 2, patch layer 2, dielectric layer 2 and metal ground. The air grid sheet composed of 10 × 10 cube units, which is located above the patch layer 1. At the same time, combined with the improved PSO optimization algorithm, the displacement x_i of the air grid of 100 cube elements is used as the optimization particle. During the optimization process, each air element can move up and down freely, and its displacement is determined by the rounding function:

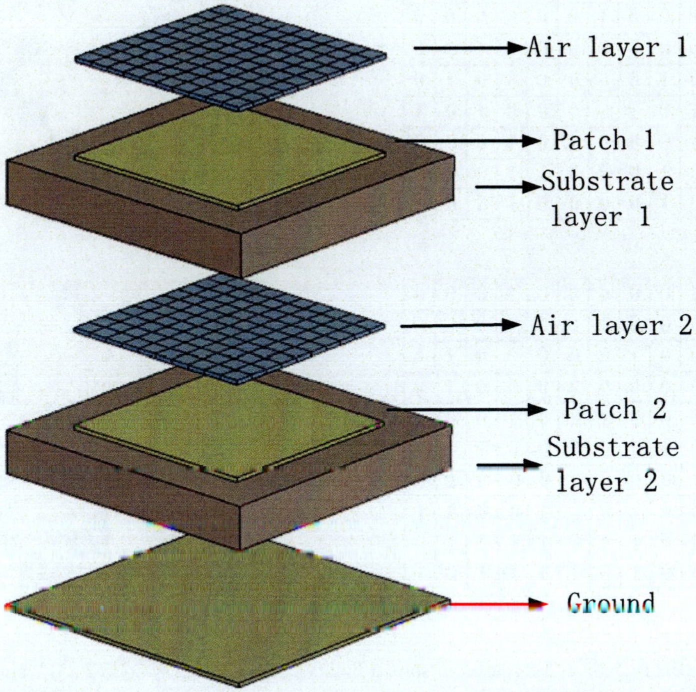

Fig. 6.1 Basic optimized model of binary coding antenna

$$\text{int}(x_i) = \begin{cases} 0 & 0 < x_i < 1 \\ 0 & 1 \leq x_i < 2 \end{cases} \quad i = 1, 2 \ldots, 100 \tag{6.1}$$

The displacement of the ith grid is

$$h = -0.035 \times \text{int}(x_i) \tag{6.2}$$

Obviously, the coding form of the grid was "1" when x_i is in $1 \leq x_i < 2$. At this time, the discrete air grid decreased by 0.035 mm to the patch layer, and then overlapped with the patch layer. By Boolean subtraction, a square space is formed in the patch layer. Similarly, the grid is coded as "0" when x_i is in $0 < x_i < 1$, and the corresponding displacement of the grid is 0, the patch layer will not change, and the patch layer still retains metal. That is, when the grid code is "0", it represents metal, and when the grid code is "1", it represents square space. Different values of random variables correspond to different random binary codes, that is, different patch structures, as shown in Fig. 6.2. The specific values and changes of random variables need to be determined according to the optimization objectives. Generally,

1	0	1	0	0	0	1	0	0	1
0	1	0	0	0	1	0	1	0	1
0	0	1	0	0	0	1	0	0	0
0	0	0	1	0	0	0	0	0	0
0	0	0	0	0	0	0	1	0	0
0	1	0	0	0	0	0	0	1	0
1	0	1	0	0	0	0	1	0	1
1	0	0	0	1	0	1	0	0	0
0	1	0	0	0	1	1	0	0	0
0	1	0	0	0	0	1	0	0	1

→

(a)

1	0	0	0	0	0	1	0	0	0
0	0	0	0	0	1	0	0	0	0
0	0	1	0	0	0	0	0	1	1
0	0	0	0	0	0	0	0	1	0
0	0	0	1	0	0	1	0	0	1
0	1	0	0	0	0	0	1	1	1
1	0	0	0	1	0	0	0	1	0
0	0	0	0	0	1	0	0	1	1
0	0	1	1	0	1	1	0	1	0
0	0	1	0	1	0	1	1	0	1

→

(b)

Fig. 6.2 Binary code and it represent state **a** binary code 1, **b** binary code 2 ("0" represent metal, "1" represent substrate, "yellow" represent metal, and "black" represent substrate)

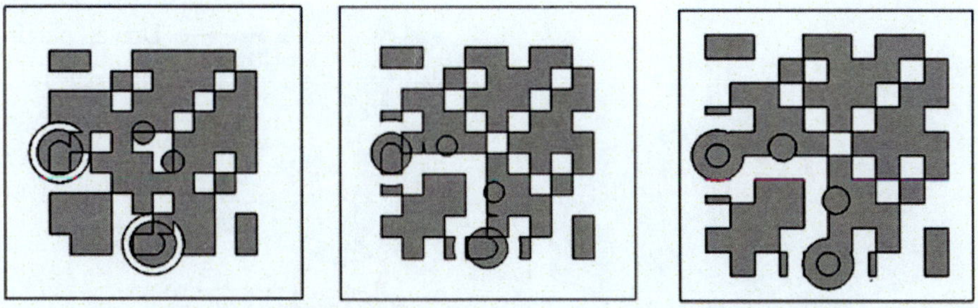

Fig. 6.3 Different optimized model of binary coding antenna

the S parameters and pattern of the antenna are set as the optimization objective function.

Figures 6.2 and 6.3 are only the coding and structure of some intermediate states during the optimization process of millimeter wave antenna. It can be seen that when the coding is different, the grid structure of the antenna is different. Accordingly, when the antenna is excited, the distribution of surface current, electric field and magnetic field are also different, and the radiation will be different. Such a binary coded grid antenna model has a large degree of freedom and can realize the target design without relying on human experience.

The final optimized millimeter wave antenna structure is shown in Fig. 6.4. It consists of two dielectric layers and three copper film layers, in which the dielectric layers are Arlon250 dielectric layers with a thickness of 1.524 mm and a relative dielectric constant of 2.5. And the three copper films are connected through a through hole with a radius of r_1. The bottom copper film is the ground, the middle copper film is the parasitic layer, and the top copper film is the driven patch. The driven patch is fed by two symmetrical coaxial probes, which are connected to the SMA under the ground. In order to facilitate the welding between the coaxial probe and the driving patch, two rings with radius r_2 are introduced.

According to the optimization construction method mentioned above, the parasitic layer and driven patch are also the optimized binary coded grid structure, as shown in Fig. 6.4c, d. Since two ports excite two orthogonal polarization EM signals, the structure is diagonally symmetrical during initialization. In addition, in the specific optimization design process, by adopting the finite integral technology (FIT) and particle swarm optimization algorithm, the combination of the impedance bandwidth, upward radiation and beam width is set as multi-objective optimization functions in CST Microwave Studio software, and the x_i sample on each grid is optimized.

6.3 Geometry Structure of Antenna

The final optimization parameters of binary coded millimeter wave dual polarization antenna structure shown in Fig. 6.4 are shown in Table 6.1.

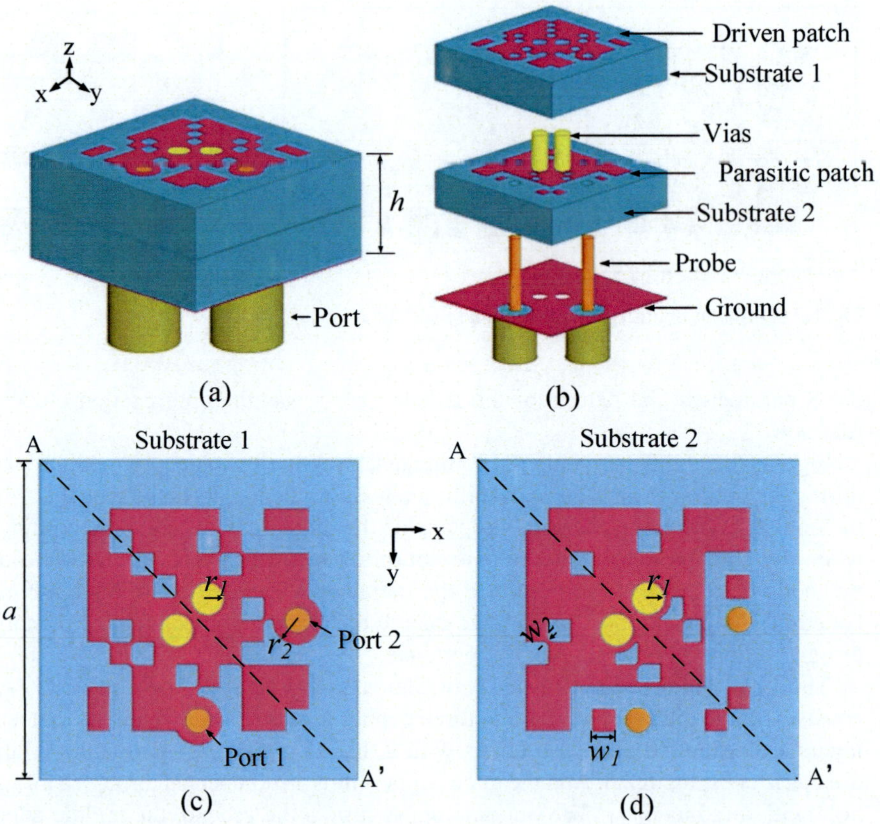

Fig. 6.4 Configuration of proposed antenna. **a** 3-D view of proposed antenna, **b** exploded view showing individual layers, **c** driven patch, **d** parasitic patch. Figure reproduced with permission from Ref. [18], © 2020 IEEE

Table 6.1 Optimized geometric parameters of the proposed antenna

Parameters	w_1	w_2	a	r_1	r_2	h
Values (mm)	0.46	0.1	6.6	0.325	0.51	3.11

Optimizing x_i samples on each mesh, some square holes can be properly introduced and modify the shape of patch. Furthermore, multiple branch sections are generated on top and parasitic patches. These branches just resonate in different frequencies. All of these resonant bands are synthesized into the final available broadband.

In order to further explain the principle of binary grid optimization antenna, we analyze its current distribution characteristics. The transient surface current distribution on the antenna driven patch and parasitic patch is shown in Fig. 6.5. Obviously, there are several current lines, and different length of current paths correspond to

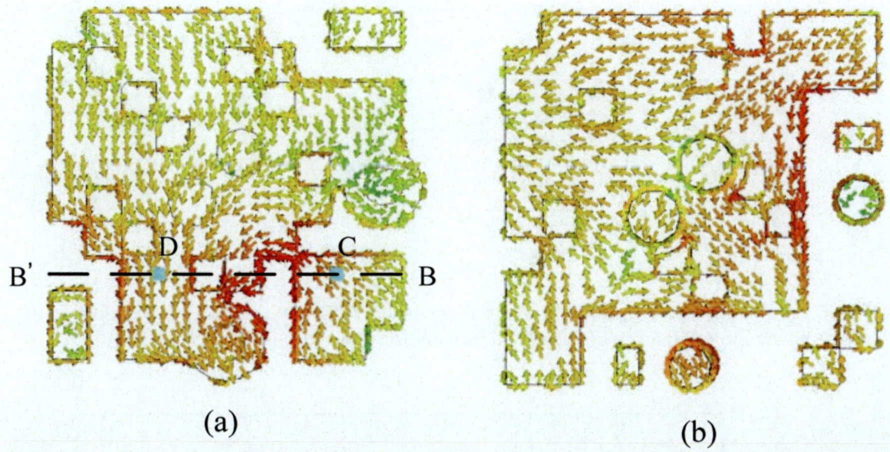

Fig. 6.5 The surface current distribution on the driven patch and parasitic patch for port 1 at 25 GHz **a** driven patch, **b** parasitic patch. Figure reproduced with permission from Ref. [18], © 2020 IEEE

different resonance frequencies theoretically [15]. For the optimized patch, take out the phase of current component along the Y direction of curve B–B′, as shown in Fig. 6.6, and the current phase along curve B–B′ changes greatly. The phase difference between point C and point D is almost 165°, which is close to the inversion. Generally speaking, two current elements with inverse phase can increase beam gain inclined to horizontal direction and widen beam width [10].

In addition, the simulated polarization results of the binary coding optimized antenna are shown in Fig. 6.7. It can be observed that the polarization in the direction of +45° is the main polarization of port 1 in the wide beam width, and the polarization in the direction of −45° is the cross polarization. However, the polarization component level in the direction of −45° is almost equal to the polarization in the direction of +45° in the cell near the Z axis, but the phase difference in this region is almost 90°, so the circular polarization is formed, as shown in Fig. 6.8. Because the

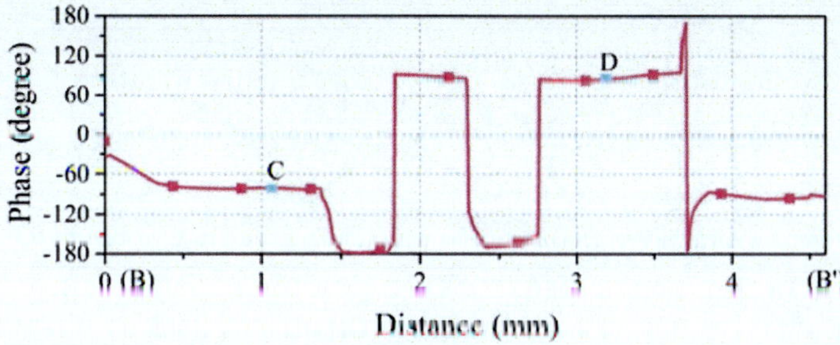

Fig. 6.6 Simulated surface current phase distribution of the proposed antenna for port 1 at 25 GHz. Figure reproduced with permission from Ref. [18], © 2020 IEEE

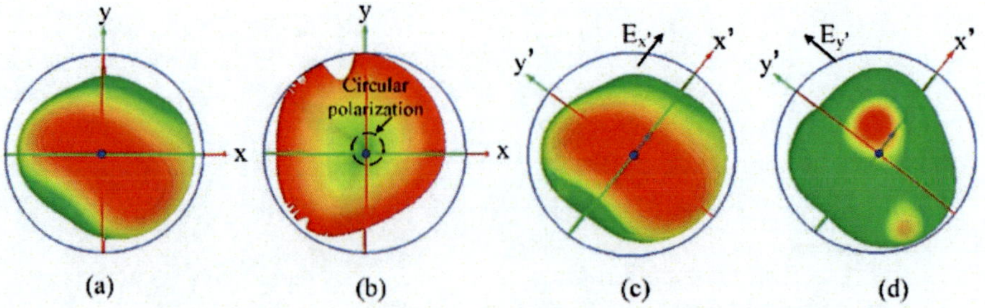

Fig. 6.7 Simulated polarization of proposed antenna **a** Abs, **b** axial ratio, **c** +45° polarization, **d** −45° polarization. Figure reproduced with permission from Ref. [18], © 2020 IEEE

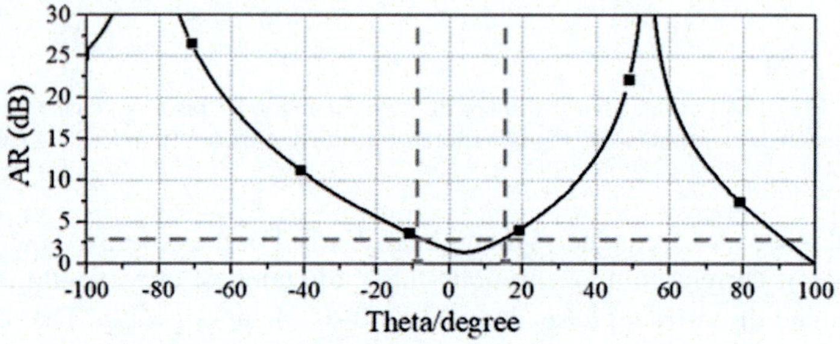

Fig. 6.8 The axial ratio of the proposed binary coding antenna

binary coding optimization antenna is symmetrical, the polarization characteristics of port 2 can be deduced too.

It is assumed that the binary coding antenna optimized in this section is Ant 1 and the traditional dual port microstrip antenna with parasitic patch is Ant 2. The simulation results of the reflection coefficient and isolation of the two ports of these two antennas are shown in Fig. 6.9. The simulated impedance bandwidth of Ant 1 is in 21.8–33.2 GHz, the absolute bandwidth is 11.4 GHz, and the relative

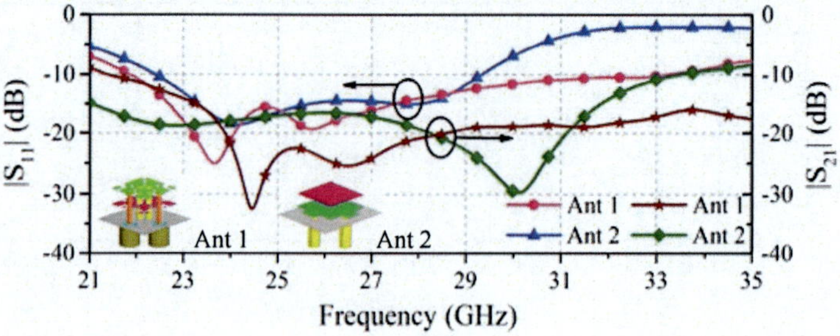

Fig. 6.9 Simulated return loss and isolation of Ant 1 and Ant 2. Figure reproduced with permission from Ref. [18], © 2020 IEEE

impedance bandwidth is 41.5%. The simulated impedance bandwidth of Ant 2 is in 22.4–29.3 GHz, and the relative impedance bandwidth is 27.7%. It can be seen that the bandwidth of Ant1 is 13.8% wider than that of Ant2. In addition, the port isolation of Ant1 is 10 dB greater than that of Ant2 within the impedance bandwidth, which shows that after introducing binary coding optimization random variables to modify the patch structure, the two ports of Ant1 maintain a very good isolation.

During the optimization to the binary coding antenna, wide beamwidth is one of the conditions for multi-objective optimization function. The simulated far fields at different frequencies are shown in Fig. 6.10, in which the pattern at f = 28 GHz is slightly offset but wide beam. At f = 30 GHz, the pattern has high gain at sides and low gain in center. These pattern characteristics can compensate for the gain roll off during large angle scanning when forming the array.

Figure 6.11 is a comparison of normalized radiation patterns in the x–O–z plane of port 1 at 25 GHz. It can be observed that the beam width of the traditional Ant 2 is only 73.4°, while the beam width of the binary coding optimized Ant 1 is as high as 141°. This also shows that Ant 1 has great potential as an antenna array element for wide-angle scanning.

6.4 Result Analysis

6.4.1 The Experiment Results of Antenna Element

According to the optimization results in the above section, we actually processed a binary coded dual polarization millimeter wave antenna, and measured the radiation characteristics of the antenna in the chamber, as shown in Fig. 6.12.

Figure 6.13 shows the S parameters of the measured antenna prototype, which basically coincides with the simulation results. The frequency range with return loss less than −10 dB is 21.8–33.2 GHz, and the relative impedance bandwidth is about 41.5%. Within the effective bandwidth, the isolation between the two ports of the antenna is large, which is roughly the same as the simulation trend.

Figure 6.14 is the measured patterns in x–O–z plane and y–O-z plane when antenna port 1 is excited, which is in good agreement with the simulation results. At f = 24 GHz, 27.5 GHz and 30 GHz, the half power beam widths in the x-O-z plane are 154°, 128° and 149° respectively. The radiation pattern is relatively stable and the beam is wide.

Figure 6.15 is the measured gain curve of antenna element. The gain in the whole impedance bandwidth is basically greater than 4 dB, up to 5.8 dB at most. Compared with the simulation results, the measured gain curve of the antenna fluctuates around the simulation curve, which may be caused by the machining accuracy of the millimeter wave antenna and the alignment of the reference line during the test.

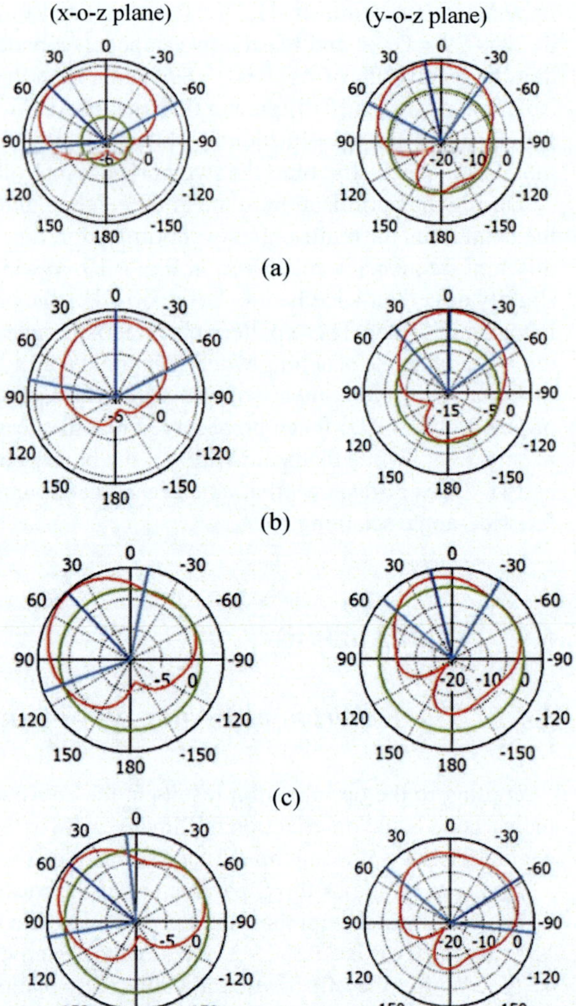

Fig. 6.10 The simulated radiation pattern in x–o–z plane and y–o–z plane. **a** 24 GHz, **b** 26 GHz, **c** 28 GHz, **d** 30 GHz, **e** 32 GHz

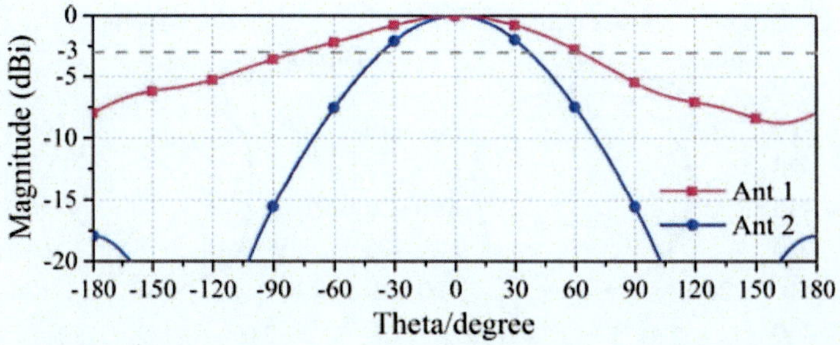

Fig. 6.11 Simulated radiation pattern of Ant 1 and Ant 2 for port 1 in x–o–z plane at 25 GHz. Figure reproduced with permission from Ref. [18], © 2020 IEEE

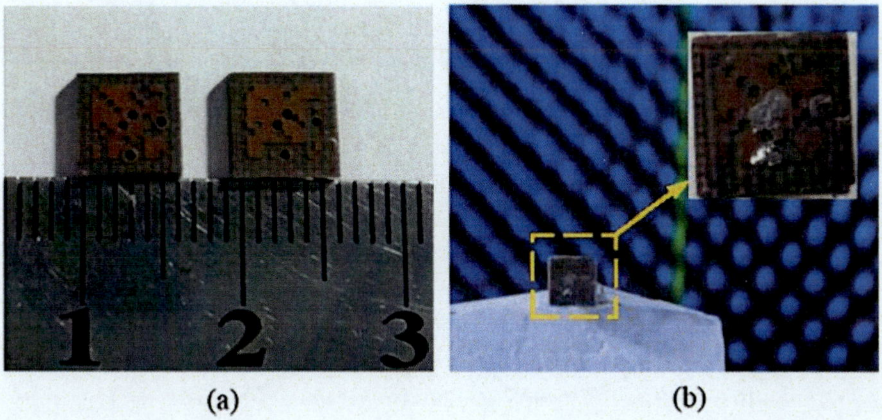

Fig. 6.12 Simulated radiation pattern of Ant 1 and Ant 2. **a** The view of different layers, **b** measurement environment

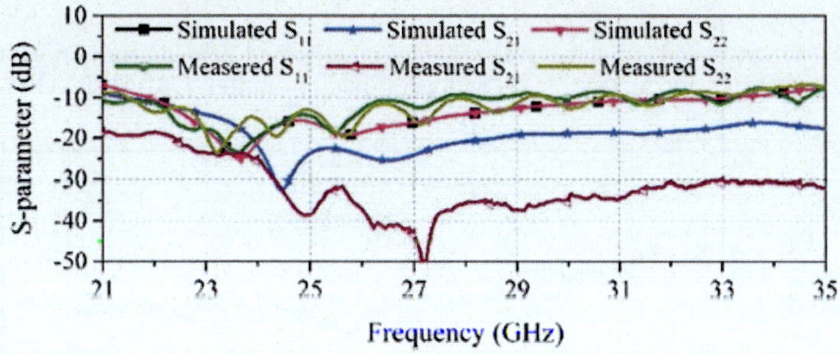

Fig. 6.13 Measured and simulated S-parameters of proposed antenna. Figure reproduced with permission from Ref. [18], © 2020 IEEE

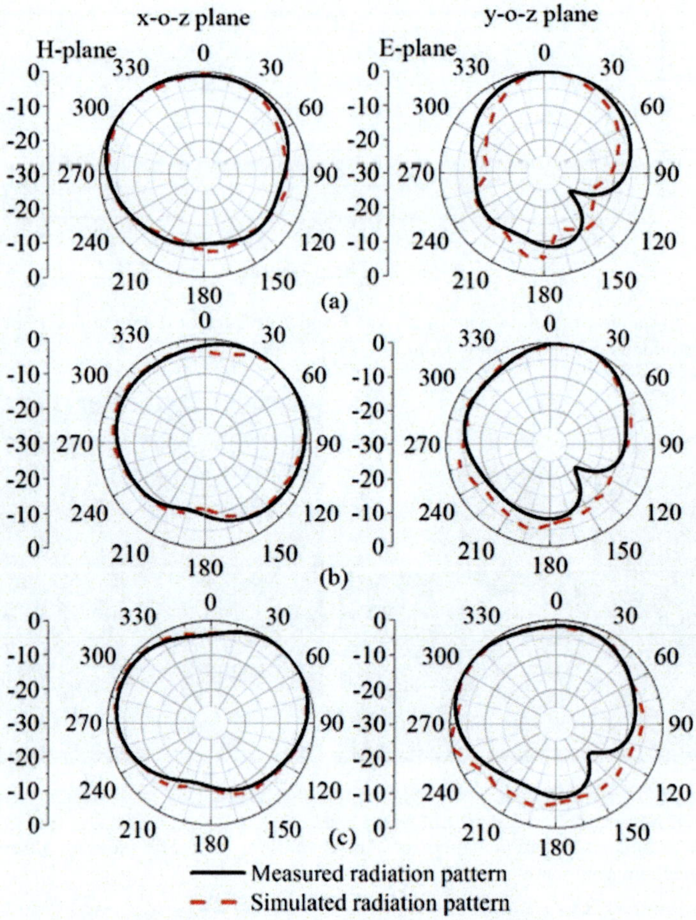

Fig. 6.14 Measured and simulated realized gain of proposed antenna for port 1. **a** 24 GHz, **b** 27.5 GHz, **c** 30 GHz. Figure reproduced with permission from Ref. [18], © 2020 IEEE

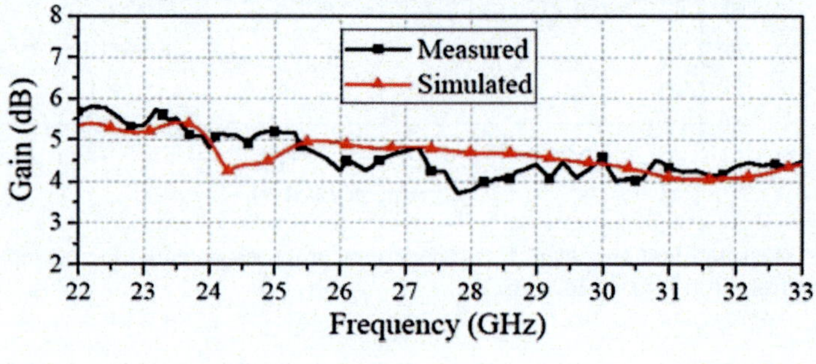

Fig. 6.15 Measured and simulated realized gain of proposed antenna. Figure reproduced with permission from Ref. [18], © 2020 IEEE

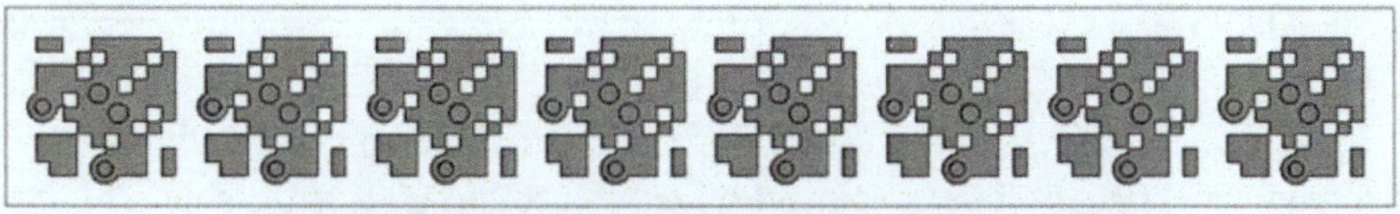
(a)

(b)

Fig. 6.16 The prototype of proposed antenna array in the anechoic chamber. **a** The simulated model, **b** The measured prototype. Figure reproduced with permission from Ref. [18], © 2020 IEEE

6.4.2 *The Experiment Results of Array*

According to the performance of the previous optimized antenna element, the dual port antenna based on binary coding has good broadband and wide beam performance. Based on the principle of pattern product, the binary coded millimeter wave dual polarization antenna can be extended into 1 × 8 array to realize large angle scanning.

Expand the elements along the x-axis to achieve a 1 × 8 linear array. The array element spacing is less than half wavelength, and the array element spacing is 4.6 mm based on the central frequency point of 27.5 GHz. According to the antenna array model, the antenna array is actually manufactured and measured in the chamber, as shown in Fig. 6.16.

The array scanning experiment results are shown in Fig. 6.17, which are basically consistent with the simulation results. When f = 24 GHz, the scanning range can reach −65° to + 65°, and the gain roll off is less than 3 dB. When f = 27.5 GHz, the measured scanning angle range of the antenna is −50° to +50°, the antenna gain roll off reaches 4.1 dB at + 50°, but the gain roll off is only 1.4db at −50°, and the antenna gain roll off is less than 3 dB at −60°. The scanning effect at f = 30 GHz is shown in Fig. 6.17c. The gain roll off of the array in the scanning range of 100° is less than 3 dB. The gain range of the antenna is about 9.3 to 12.1 dBi @ 24 GHz, about 9.2–12.1 dBi @ 27.5 GHz and 9–11.4 dBi @ 30 GHz. In short, the millimeter wave antenna array has good large angle scanning ability.

Table 6.2 shows the comparison between the proposed binary coded antenna (array) and other reported antennas. Tthe proposed antenna can work in wide bandwidth of mininium meter band with dual polarization, and wide beamwidth. And the phased array composed of the proposed antenna prototype can scan large angle.

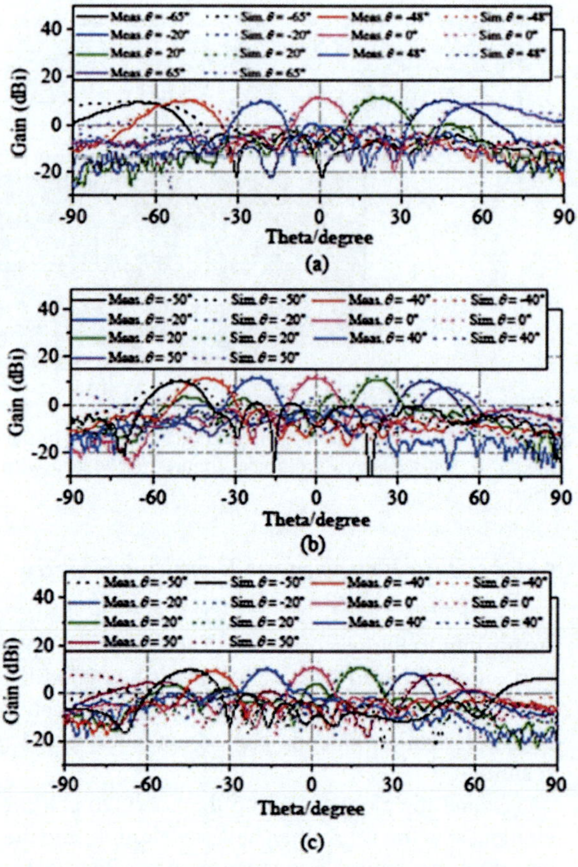

Fig. 6.17 Comparison of the simulated and measured beam scanning of proposed phased array in x–o–z plane for port 1. **a** 24 GHz, **b** 27.5 GHz, **c** 30 GHz. Figure reproduced with permission from Ref. [18], © 2020 IEEE

Table 6.2 Comparisons between the proposed antenna and other models [18], © 2020 IEEE

References	BW (%)	Polarization	Beamwidth	Beam-scanning angle (array)
[15]	16	Dual polarization	100°	/
[10]	37	Single polarization	/	≥40°
[13]	23.5	Single polarization	/	/
[16]	12.6	Single polarization	102°	/
[17]	9.9	Single polarization	116°	/
This work [18]	41.5	Dual polarization	128°	≥100°

6.5 Summary

A broadband dual-polarized binary coding antenna with wide beamwidth in mm-wave band is proposed. The antenna array element adopts a double-layer structure, the parasitic patch and driving patch are optimized in the form of binary coding, so as to finally realize the characteristics of broadband and wide beam. Based on the above characteristics, a large angle scanning antenna array is designed and implemented. The measured results show that the impedance bandwidth of the antenna unit is 21.8–33.2 GHz, the relative impedance bandwidth is about 41.5%, the half power beam width is 128°–157°, and gain 4–5.8 dBi. The extended array has the characteristics of large angle scanning, and the beam scanning range is greater than 100°. The designed antenna has a good application prospect in 5G millimeter wave communication.

References

1. T.S. Rappaport et al., Millimeter wave mobile communications for 5G cellular: It will work! IEEE Access **1**, 335–349 (2013)
2. J.G. Andrews et al., What will 5G be? IEEE J. Sel. Areas Commun. **32**(6), 1065–1082 (2014)
3. J. Qiao, X.S. Shen, J.W. Mark, Q. Shen, Y. He, L. Lei, Enabling device-to-device communications in millimeter-wave 5G cellular networks. IEEE Commun. Mag. **53**(1), 209–215 (2015)
4. Y. Chen, S. De, R. Kernchen, K. Moessner, Device discovery in future service platforms through SIP, in *2012 IEEE Vehicular Technology Conference (VTC Fall)*, Quebec City, QC, pp. 1–5 (2012)
5. Y. Azar et al., 28 GHz propagation measurements for outdoor cellular communications using steerable beam antennas in New York city, in *2013 IEEE International Conference on Communications (ICC)*, Budapest, pp. 5143–5147 (2013)
6. Z. Qingling, J. Li, Rain attenuation in millimeter wave ranges, in *2006 7th International Symposium on Antennas, Propagation & EM Theory*, Guilin, pp. 1–4 (2006)
7. T.S. Rappaport, J.N. Murdock, F. Gutierrez, State of the art in 60-GHz integrated circuits and systems for wireless communications. Proc. IEEE **99**(8), 1390–1436 (2011)
8. R.B. Waterhouse, D. Novak, A. Nirmalathas, C. Lim, Broadband printed millimeter-wave antennas. IEEE Trans. Antennas Propag. **51**(9), 2492–2495 (2003)
9. J. Kornprobst, K. Wang, G. Hamberger, T.F. Eibert, A mm-wave patch antenna with broad bandwidth and a wide angular range. IEEE Trans. Antennas Propag. **65**(8), 4293–4298 (2017)
10. C. Mao, S. Gao, Y. Wang, Broadband high-gain beam-scanning antenna array for millimeter-wave applications. IEEE Trans. Antennas Propag. **65**(9), 4864–4868 (2017)
11. K. Gulur Sadananda, M.P. Abegaonkar, S.K. Koul, Gain equalized shared-aperture antenna using dual-polarized ZIM for mm wave 5G base stations. IEEE Antennas Wirel. Propag. Lett. **18**(6), 1100–1104
12. J. Xu, H. Ke, Y. He, Y. Luo, A wideband u-slot microstrip patch antenna for large-angle mmw beam scanning, in *2018 IEEE International Conference on Computer and Communication Engineering Technology (CCET)*, pp. 142–145 (2018)
13. W. He, R. Jin, J. Geng, E-shape patch with wideband and circular polarization for millimeter-wave communication. IEEE Trans. Antennas Propag. **56**(3), 893–895 (2008)
14. B.Q. Wu, K. Luk, A broadband dual-polarized magneto-electric dipole antenna with simple feeds. IEEE Antennas Wirel. Propag. Lett. **8**, 60–63 (2009)
15. L. Schulwitz, A. Mortazawi, Millimeter-wave dual polarized lshaped horn antenna for wide-angle phased arrays. IEEE Trans. Antennas Propag. **54**(9), 2663–2668 (2006)

16. K. Ng, C.H. Chan, K. Luk, Low-cost vertical patch antenna with wide axial-ratio beamwidth for handheld satellite communications terminals. IEEE Trans. Antennas Propag. **63**(4), 1417–1424 (2015)
17. X. Chen, P. Qin, Y.J. Guo, G. Fu, Low-profile and wide-beamwidth dual-polarized distributed microstrip antenna. IEEE Access **5**, 2272–2280 (2017)
18. L. Wang, J.-P. Geng, K. Wang, H. Zhou, C. Ren, H. Wu, X. Zhao, C. He, X. Liang, W. Zhu, R. Jin, Wideband dual-polarized binary coding antenna with wide beamwidth and its array for millimeter-wave applications. IEEE Antennas Wirel. Propag. Lett. 19(4x), 636–640 (2020)

Chapter 7
Design UWB Antenna by Binary Coding Optimization Method

7.1 Introduction

Inspired by the development and evolution of nature, many natural laws can also be used to solve antenna design problems, such as artificial neural network technology, genetic algorithm, simulated annealing algorithm, particle swarm optimization algorithm, etc. [1] many optimization algorithms are widely used in antenna design based on the varying laws and mathematical models of nature.

With the rise of UWB transmission technology, the design of UWB antenna needs to consider the requirements of UWB, phase linearization, pattern uniformity and waveform fidelity. The design of UWB antenna using intelligent optimization algorithm is essentially a multi-index and multi-objective optimization problem. The traditional optimization design of UWB antenna mostly depends on the experience of designers, especially the selection of the basic geometry structure of the antenna, which also makes it difficult to realize the automatic design of UWB antenna by computer, resulting in the limited tunable range and design performance of the antenna. The optimization design of UWB antenna based on multi-objective function can avoid many human factors (such as the limitation of experience), and realize the performance optimization on the basis of making full use of the effective size.

Genetic algorithm (GA), particle swarm optimization (PSO) algorithm and simulated annealing algorithm are commonly used to optimize UWB antenna. References [2–4] combined genetic algorithm (GA) and finite difference time domain (FDTD) method to optimize multiple parameters of UWB antenna, realized the UWB antenna, achieved low standing wave ratio, realized the correlation between input / output signals in time domain, and the antenna dispersion was small in the effective frequency band. In reference [5], particle swarm optimization algorithm is used to optimize UWB antenna, and the impedance bandwidth is 3.7–9.2 GHz. Reference [6] used simulated annealing optimization algorithm and finite element method to design UWB antenna, and the impedance bandwidth reached 5 octaves. The above antenna optimization design is to optimize the structural parameters to obtain the final structure after the designer gives the preliminary antenna prototype according

J. Geng and R. Jin, *Antenna Optimization and Design Based on Binary Coding*, Modern Antenna, https://doi.org/10.1007/978-981-16-7965-0_7

to experience. Conversely, given multiple boundary condition constraints, especially miniaturization, low profile, directional radiation, UWB and high gain, it is difficult to design a target antenna that meets all conditions. Therefore, we urgently need to reduce artificial a priori intelligent antenna design as much as possible [7, 8]. References [9, 10] used genetic algorithm to automatically design UWB antenna with stable performance.

This chapter aims to design the UWB antenna that meets multiple performance indexes, introduces the binary coded grid structure, uses the combination of genetic algorithm and finite-difference time-domain electromagnetic calculation method to optimize the multi-objective parameters, and automatically designs the UWB antenna and band stop UWB antenna. In addition, binary particle swarm optimization algorithm is introduced to improve the efficiency of optimization design. Finally, the correctness and feasibility of the rapid automation design method are verified by comparing with the simulation results of commercial software and experimental tests.

7.2 UWB Antenna Parameter Optimization Based on Traditional Method

Firstly, the fitness function must be selected to optimization process. The same problem can be optimized by different fitness of the function. What kind of fitness function to use also needs a deep understanding of the optimization problem and boundary conditions.

7.2.1 Selection of Fitness Function

For example, for the general traditional antenna, the impedance bandwidth is the concern of the initial designer, so the designer usually takes the maximum value of the return loss S_{11} parameter of the antenna feeding port within the target frequency band as the fitness function. Generally, the S_{11} values of several points are extracted as the sampling point within the working frequency band, and then the maximum value of S_{11} is obtained

$$G = \max S_{11}(i) \quad (i = 1, \ldots, N) \tag{7.1}$$

where G is the fitness function and N is the total number of sampling points.

In the literature, the average value of S11 in the target band is set as the fitness function:

$$G = \frac{1}{N}\sum_{i=1}^{N} S_{11}(i) \tag{7.2}$$

In addition, considering that the S11 parameter may fluctuate greatly in the UWB, and even a small local value will squeeze the values of other frequency points, resulting in large S11 parameters at some frequency points, the fitness function must be adjusted accordingly. Here, we compromise and take −15 dB value of S_{11} as the standard to measure the fitness function to optimize the UWB characteristics. The fitness function is expressed as:

$$\begin{aligned} &fitness = \frac{1}{N}\sum_{i=1}^{N} S_{11}(i) \\ &where\ S_{11}(\mathrm{dB}) = \begin{cases} S_{11}(\mathrm{dB}) + 15\ \mathrm{dB}, & S_{11}(\mathrm{dB}) \geq -15\ \mathrm{dB} \\ 0, & S_{11}(\mathrm{dB}) < -15\ \mathrm{dB} \end{cases} \end{aligned} \tag{7.3}$$

In the application, according to the American FCC standard, the optimized frequency band is set as 3–11 GHz, the sampling interval is 400 MHz, a total of 21 sampling points, and the arithmetic mean is taken as the fitness function.

7.2.2 Structural Parameter Optimization Process

Firstly, we use the traditional optimization method to optimize an UWB antenna. A common stepped patch antenna is selected as the initial prototype, and the step cutting shape is used to approximate the UWB antenna.

The initial structure of the antenna is shown in Fig. 7.1, the dielectric constant of the medium is 2.65 and the overall size is 23.32 mm × 34.45 mm × 1.06 mm, patch size is 23.32 mm × 19.08 mm.

The optimized structural parameters are W1, W2, W3, W4, L1, L2, L3 and L4, which correspond to the width and length of the small rectangle that can be cut off on both sides of the patch. These eight parameters are compiled into binary codes, which describe 8 the shape of the antenna. When compiling binary code, W1, W2, W3 and W4 are all set as 4-bit binary numbers, L1, L2, L3 and L4 are all set as 5-bit binary numbers, and the total length of the individual is 36 bit binary numbers. Using the fitness function of Eq. (7.3), the fitness function value is obtained after each FDTD calculation, and the decision of individual inheritance or variation is given from the fitness value, and then a new generation of individuals is generated through selection, crossover and variation. The above process is repeated until the set objective function requirements are met or the set genetic algebra is reached, and the optimization process is terminated to obtain the optimization results.

The flow chart of optimization design based on genetic algorithm [24–27] and finite difference time domain method [22, 23] is shown in Fig. 7.2.

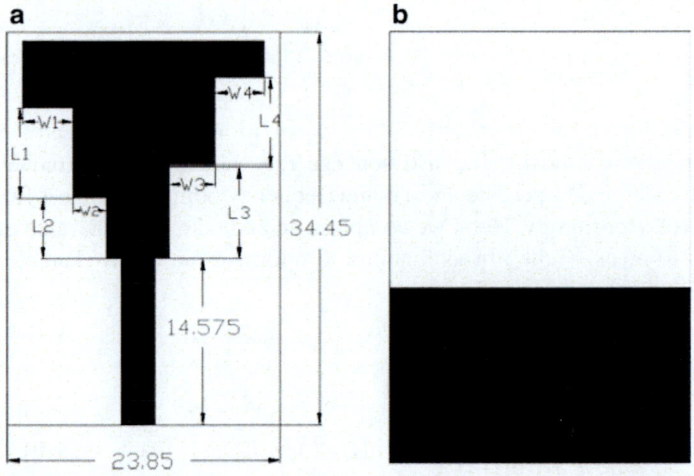

Fig. 7.1 The original structure optimized **a** the antenna element, **b** the ground [11]

In the process of optimization design, the step size in X direction, Y direction and Z direction calculated by FDTD is 0.265 mm (Δ). The thickness of the medium is 3 Δ, and the width of microstrip line is 8 Δ. The total number of grids is 130 × one hundred and twenty × 37。The antenna is fed by a unit amplitude Gaussian pulse at the end of the microstrip line. The time of each chromosome was 6000 Δ t.

7.2.3 Optimization Results

The optimized antenna is shown in Fig. 7.3. And the size parameters are: W1 = 3.71 mm, W2 = 2.12 mm, W3 = 0.53 mm, W4 = 0.795 mm, L1 = 7.155 mm, L2 = 2.12 mm, L3 = 5.035 mm, L4 = 5.83 mm.

The return loss curve of the antenna calculated with FDTD is shown in Fig. 7.4, which is close to the result calculated by commercial software CST. And the impedance bandwidth covers the frequency band of 3-11 GHz. In addition, the simulated phase stability of the optimized antenna within the impedance bandwidth is shown in Fig. 7.5. It fluctuates slightly around 0°, which is relatively stable.

The far-field patterns of the antenna at 3, 7 and 10 GHz are shown in Fig. 7.6. The pattern of the antenna deflects at different frequencies, mainly because the antenna structure is asymmetric.

The gain versus frequency curve of the optimized antenna is shown in Fig. 7.7. In the impedance bandwidth, the gain is stable in the range of 2.59–4.76dBi.

It can be seen from the above calculation results that an ultra wideband antenna is optimized and designed based on the existing a priori structure based on the combination of genetic algorithm and FDTD, but the designer's experience and imagination space limit the design of the antenna structure.

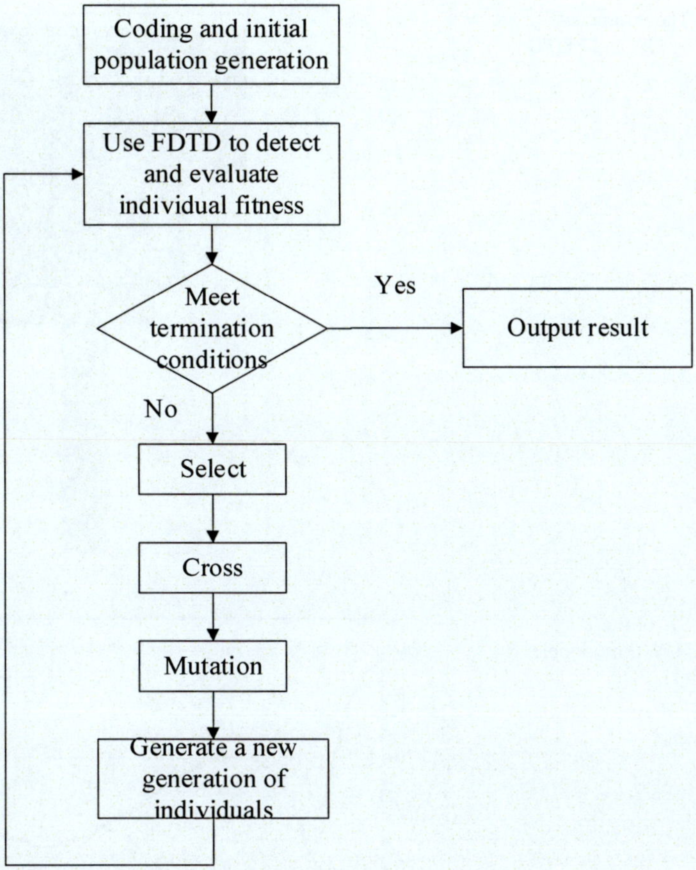

Fig. 7.2 The flow chart of GA and FDTD

7.3 Automatic Design of UWB Antenna Based on GA and FDTD

Obviously, to find the optimal structure of UWB antenna, it is not enough to optimize the structure size alone, especially the prototype structure based on the designer's experience limits the improvement space of antenna performance. On the contrary, using the binary coded grid structure prototype described in Chap. 2, the codes "1" and "0" represent metal or space respectively, which is possible to approximate any two-dimensional structure and design the antenna with the best shape under strong constraints.

2D genetic algorithm is used to optimize UWB patch antenna. The basic model of the antenna is constructed according to the grid method of binary coding. Each grid is a basic unit, and each unit corresponds to a binary number in the two-dimensional chromosome.

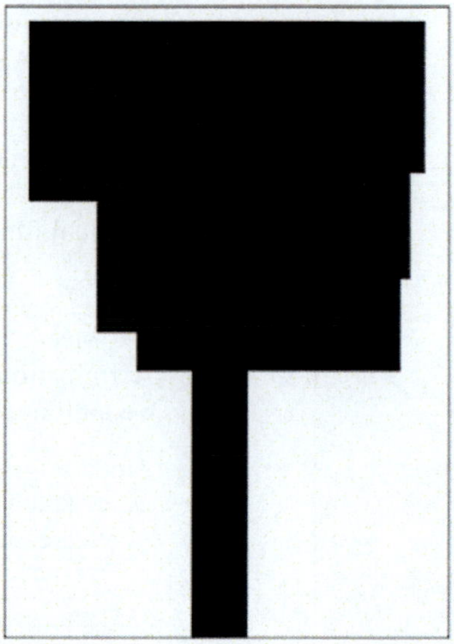

Fig. 7.3 The optimized structure by GA and FDTD [11]

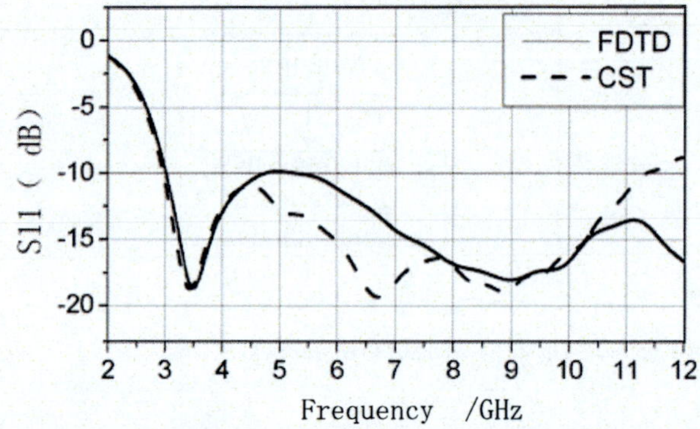

Fig. 7.4 S_{11} of the optimized UWB monopole antenna [11]

At the same time, the finite difference time domain (FDTD) method [22, 23] is used to calculate the electromagnetic wave propagation and radiation field distribution after antenna feeding, and calculate the return loss S11 of the antenna according to the electromagnetic response characteristics in time domain. Then the fitness function is calculated according to the electromagnetic characteristics of the current model of the antenna.

In the operation process of genetic algorithm, crossover is a very important step. Not only good samples should be crossed, but also good and bad samples should be crossed each other too, which can avoid premature [12]. Using the combination of the above optimization and calculation methods, a UWB printed antenna design program is achived, and a binary coded UWB antenna with a bandwidth of 3.1–12 GHz is designed.

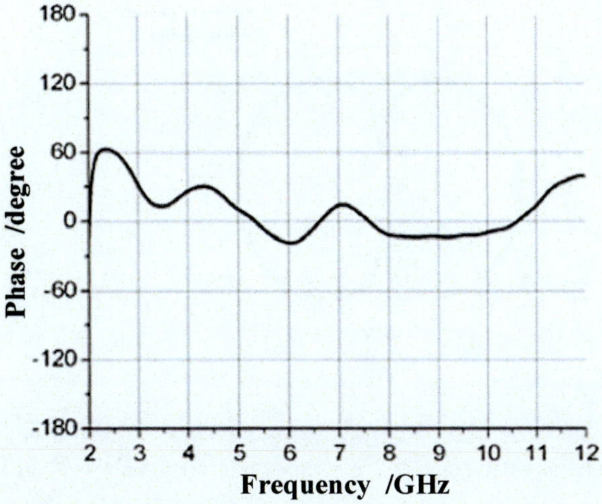

Fig. 7.5 The phase characteristic of the optimized UWB monopole antenna [11]

7.3.1 Automatic Structure Design Process

The initial size is selected according to the lowest frequency. The rectangular area is about one wavelength, considering the feeding, it is divided into 10 × 8 units, which corresponds to 10 × 8 binary chromosomes units in genetic algorithm. "1" represents the part of the unit covered by metal, and "0" represents the part of the unit not covered by metal (as shown in Fig. 7.8). The initial value of each unit is given by the random function (random distribution of "0" and "1"), and the population size is 100.

In addition, considering that each element is a rectangular grid, if the two diagonal rectangular elements are just metal, there will be singularity of point connection, which can not determine the electrical connection from the engineering practice, and is not conducive to the actual processing. Therefore, in the design, a grid is extended between the two elements to ensure the overlapping area and reduce the reflection of current at the peak. See Fig. 7.8 for details. In this way, the size of each unit will be extended. In addition, in order to ensure the feeding, the units connected with the microstrip feeder in Fig. 7.8 are set to "1", that is, metal. Thirdly, to avoid the discontinuous structure of current, that is, if "0" is surrounded by "1", then "0" is replaced by "1", and if "1" is surrounded by "0", then "1" is replaced by "0".

Being similar to the previous section, the fitness function is taken as:

$$fitness = \frac{1}{N}\sum_{i=1}^{N} S_{11}(i)$$

$$where\ S_{11}(\mathrm{dB}) = \begin{cases} S_{11}(\mathrm{dB}) + 15\ \mathrm{dB}, & S_{11}(\mathrm{dB}) \geq -15\ \mathrm{dB} \\ 0, & S_{11}(\mathrm{dB}) < -15\ \mathrm{dB} \end{cases} \tag{7.4}$$

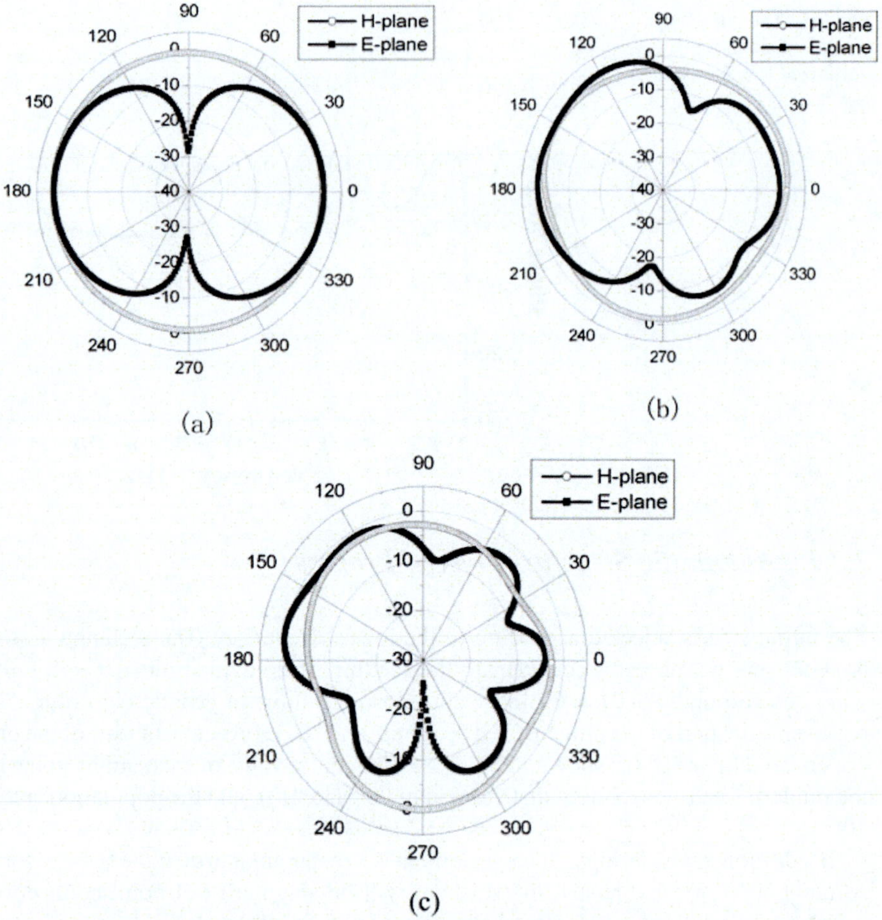

Fig. 7.6 Radiation patterns of the optimized UWB antenna **a** 3 GHz, **b** 7 GHz, **c** 11 GHz [11]

According to the ranking of fitness function values, the first 30% of samples are directly selected into the next generation, and the crossover probability is 80%. The above process is repeated until the final fitness function value reaches the design goal. The flow chart is in Fig. 7.9 as follows:

During the FDTD analysis, the step size in x and y directions is 0.53 mm (2Δ), and the step size in z direction is 0.265 mm (Δ). The thickness of the medium is 3Δ, and the width of the microstrip line is 8Δ. The total number of grids is 130 × 120 × 37. The antenna is fed with a Gaussian pulse of unit amplitude at the end of the microstrip line. The time of each chromosome is 6000Δt.

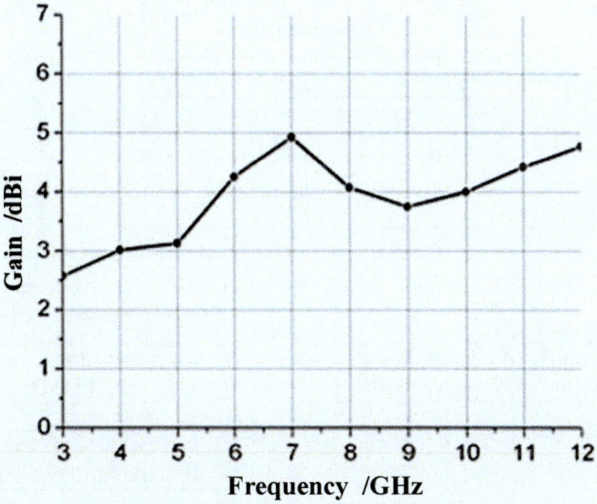

Fig. 7.7 The gain of the optimized UWB monopole antenna [11]

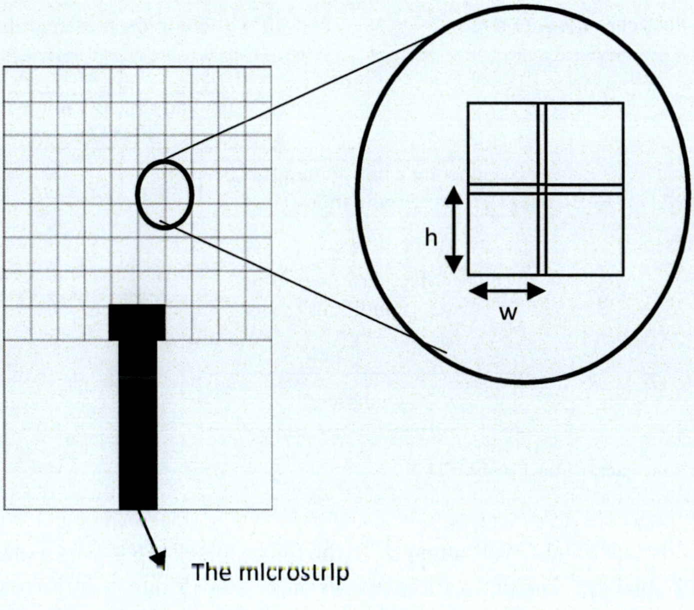

Fig. 7.8 Division of the design area with 10 × 8 grid cells

7.3.2 Design Result

After iterative calculation and optimization, the UWB antenna is finally obtained, as shown in Fig. 7.10. Figure 7.10a is the radiation unit of the antenna, and Fig. 7.10b is

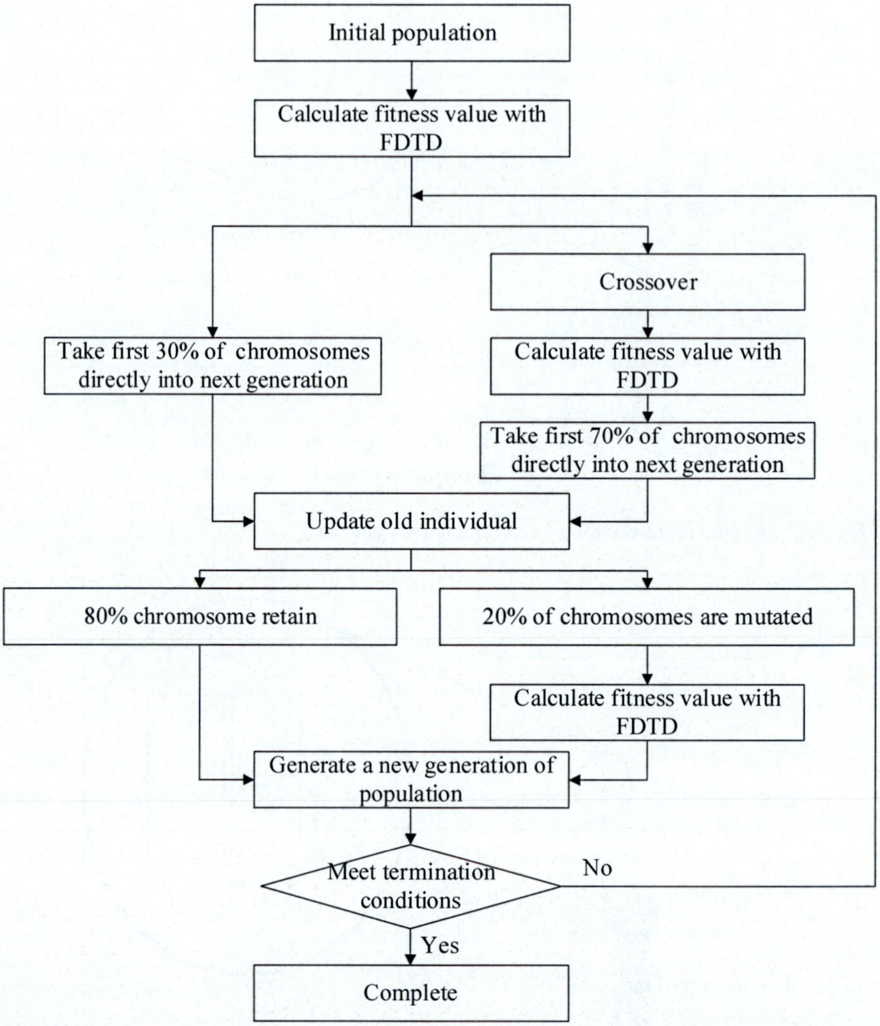

Fig. 7.9 Flow chart of the GA-FDTD

the ground plate of the UWB antenna. In the figure, black represents metal and white represents medium. The dielectric material used in the design is Teflon medium with a relative dielectric coefficient of 2.65. Figure 7.11 is a photograph of the antenna actually measured.

The S parameters of the UWB antenna prototype based on FDTD calculation, CST simulation and actual measurement are shown in Fig. 7.12. It can be seen from the figure that the available impedance bandwidth with return loss of the antenna designed by this method is less than −10 dB is more than 3.1–12 GHz. It meets the ultra wideband frequency range specified by FCC. The deviation between the

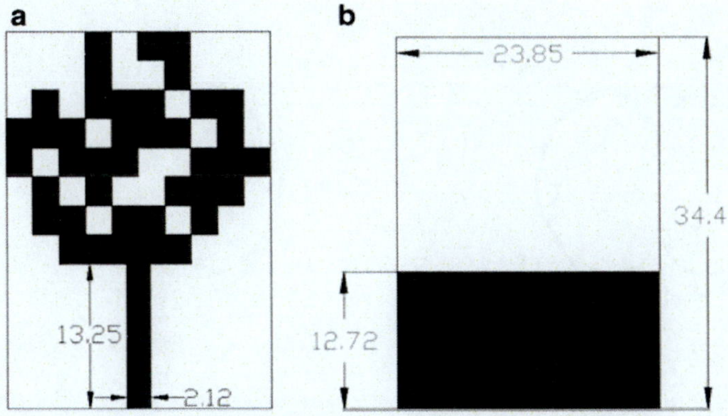

Fig. 7.10 Structure of the UWB antenna **a** the antenna element, **b** the ground. Figure reproduced with permission from Ref. [13], © 2020 John Wiley and Sons

Fig. 7.11 Photograph of the designed UWB antenna. Figure reproduced with permission from Ref. [13], © 2020 John Wiley and Sons

Fig. 7.12 Measured and simulated return losses for the designed UWB antenna. Figure reproduced with permission from Ref. [13], © 2020 John Wiley and Sons

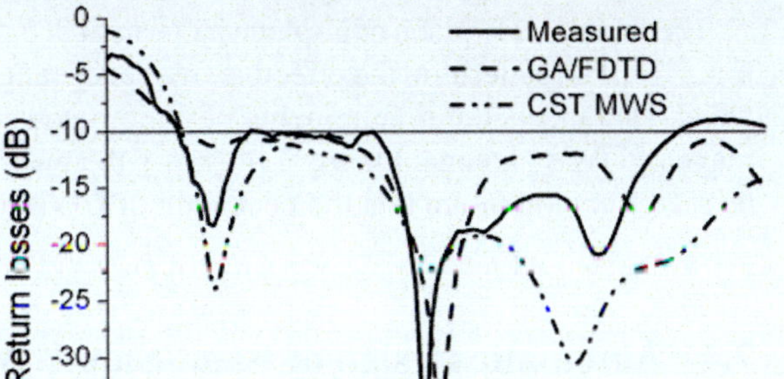

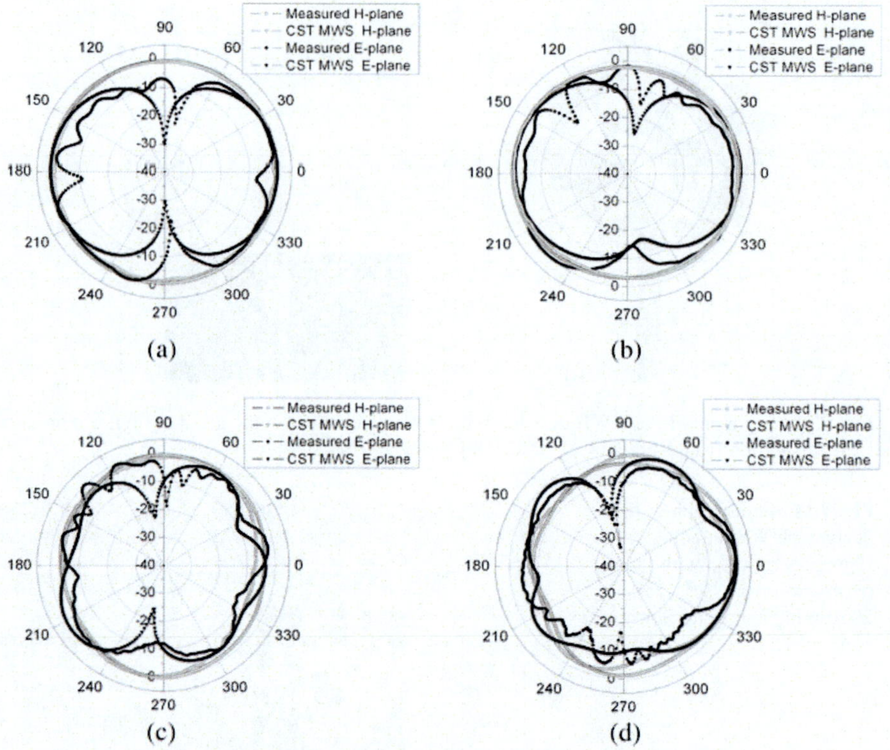

Fig. 7.13 Radiation patterns of the designed UWB antenna in different frequencies **a** 3 GHz, **b** 5 GHz, **c** 8 GHz, **d** 10 GHz. Figure reproduced with permission from Ref. [13], © 2020 John Wiley and Sons

measured results and the simulation results in the figure is mainly caused by the machining accuracy and the discontinuity between metal units.

The measured E-plane or H-plane patterns at 3, 5, 8 GHz and 10 GHz are shown in Fig. 7.13. The patterns in the effective frequency range are basically omnidirectional, which also meets the requirements of pattern consistency in the whole bandwidth range of UWB antenna. Figure 7.14 shows the measured and simulated gain. It can be seen from the figure that the peak gain of UWB antenna is 2.5–6.46dBi.

7.4 Automatic Design of Band-Stop UWB Antenna Based on GA

Considering that there are many interferences in the actual environment, especially WLAN has strong interference with UWB system at fixed frequency points, it is best to filter these interferences when receiving signals from the antenna. The previous

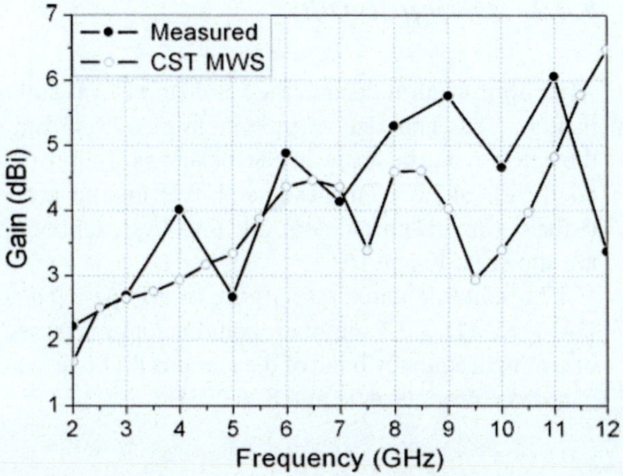

Fig. 7.14 Measured and simulated gains of the designed UWB antenna. Figure reproduced with permission from Ref. [13], © 2020 John Wiley and Sons

design mainly relied on the designer's experience to add a filter structure to the existing antenna to deduct the interference frequency band. In this section, we rely on the above GA + FDTD method and program to automatically design band stop UWB antenna within a given boundary.

7.4.1 Structural Design Process

According to the characteristics of WLAN, the stopband is 5–6 GHz. The fitness function is similar to the previous section, and the stopband range is added, as shown in Eq. (7.4).

$$S_1(f_1) = \begin{cases} S_{11}(f_1), & 2.8\text{ GHz} < f_1 < 4.8\text{ GHz}\&\&6.2\text{ GHz} < f_1 < 12\text{ GHz} \\ 0, & others \end{cases}$$

$$S_2(f_2) = \begin{cases} S_{11}(f_2), & 4.8\text{ GHz} \le f_2 \le 6.2\text{ GHz} \\ 0, & others \end{cases}$$

$$S_{1\,\text{mod}}(f_1) = \begin{cases} S_1(f_1) + 15\text{ dB}, & S_1(f_1) \ge -15\text{ dB} \\ 0, & S_1(f_1) < -15\text{ dB} \end{cases}$$

$$S_{2\,\text{mod}}(f_2) = \begin{cases} -S_2(f_2), & S_2(f_2) < -3\text{ dB} \\ 0, & S_2(f_2) \ge -3\text{ dB} \end{cases}$$

$$fitness = \frac{1}{N}\sum_{i=1}^{N}\left(S_{1\,\text{mod}}(i) + S_{2\,\text{mod}}(i)\right) \tag{7.5}$$

Other processes are the same as previous section, so we won't repeat them here.

7.4.2 Design Result

After optimization design, the band stop UWB antenna is finally designed as shown in Fig. 7.15. The size of the antenna is 23.85 mm × 34.45 mm × 0.8 mm. The dielectric material used in the design is Teflon medium with a relative dielectric coefficient of 2.65. The designed UWB antenna prototype is manufactured according to the optimized parameters, and the physical photos of this band stop UWB antenna are shown in Fig. 7.16.

The standing wave ratio of the prototype bandstop UWB antenna calculated by GA + FDTD, CST simulation and actual measurement is shown in Fig. 7.17. The operating frequency band of the antenna is above 3.1–12 GHz, and there is an obvious bandstop characteristic at 4.8–6.2 GHz.

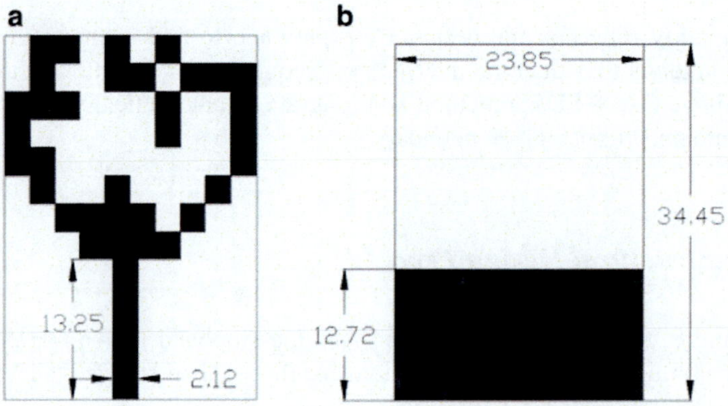

Fig. 7.15 Structure of the designed band-notched UWB antenna **a** the antenna element, **b** the ground [11]

Fig. 7.16 Photograph of the designed band-notched UWB antenna [11]

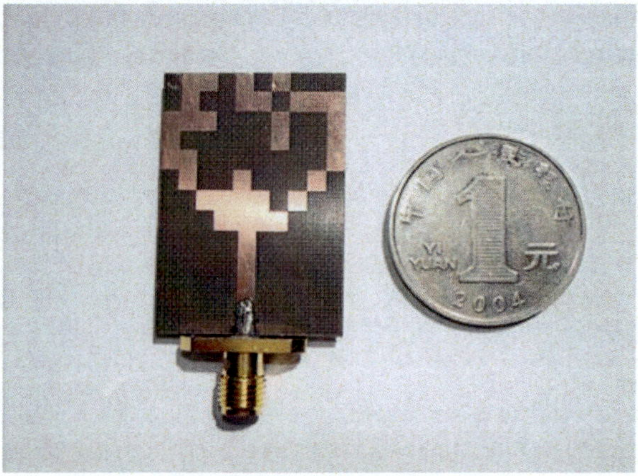

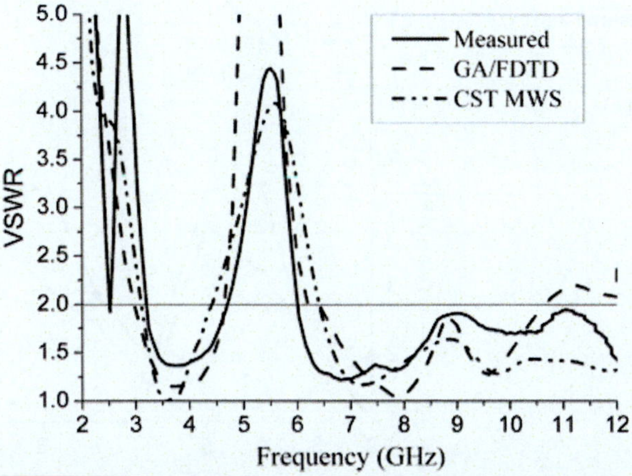

Fig. 7.17 Measured and simulated VSWR for the band-notched UWB antenna [11]

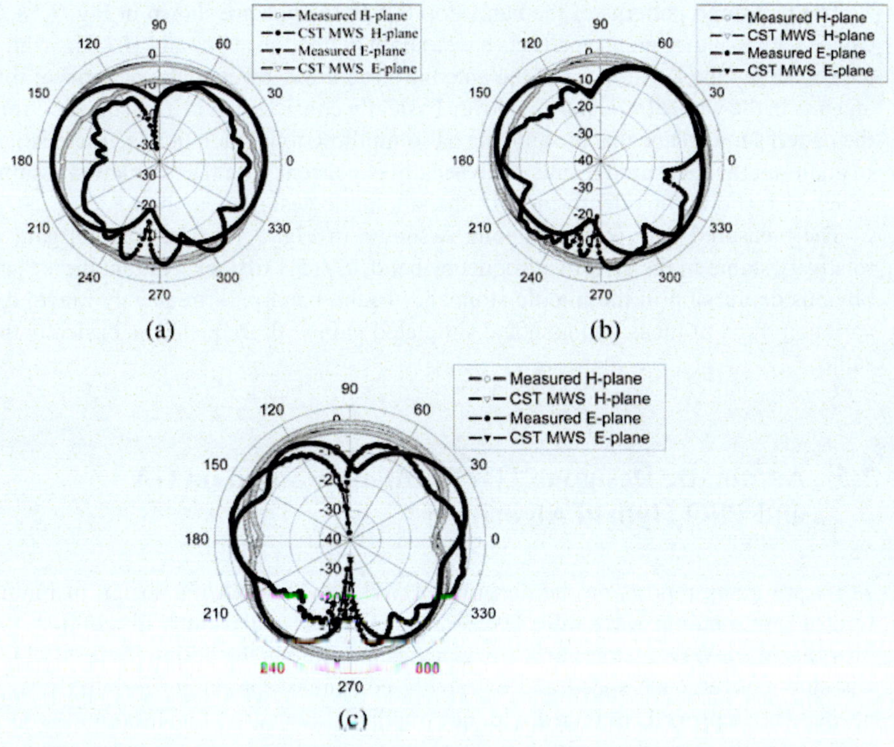

Fig. 7.18 Radiation patterns of the designed band-notched UWB antenna in different frequencies **a** 4 GHz, **b** 7 GHz, and **c** 10 GHz [11]

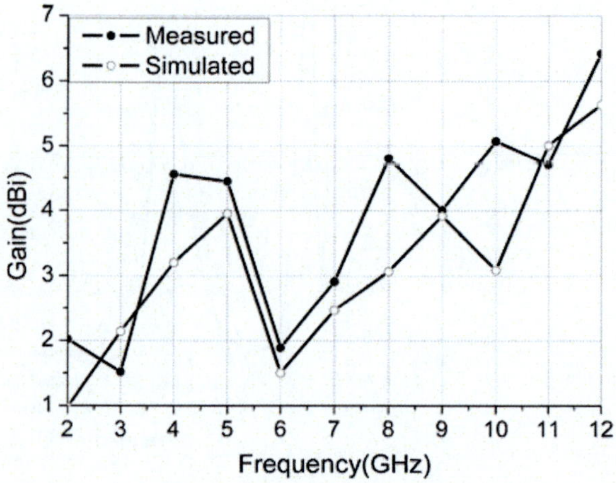

Fig. 7.19 Measured and simulated gains of the designed band-notched UWB antenna [11]

The measured patterns of the band stop UWB antenna are shown in Fig. 7.18. It can be seen that the antenna patterns havea slight deflection at 4, 7 and 10 GHz, which are caused by the asymmetry of the antenna structure. However, the patterns of this antenna in the whole frequency range are basically consistent, and it can be seen from the patterns in H plane that the antenna has omnidirectional radiation characteristics. In addition, the measured results are basically consistent with the simulation results, which also shows the effectiveness of the automatic design method.

The measured gain of the antenna is shown in Fig. 7.19. The antenna gain is relatively stable in the effective frequency band, 2.7–5.1 dBi, except that there is an obvious depression in the middle stopband. In the band-pass frequency range, the variation trend of measured gain and simulated gain with frequency is basically the same.

7.5 Automatic Design of UWB Antenna Based on GA and PSO Hybrid Algorithm

There are many reports on the design of UWB antenna, but the design of miniaturized low standing wave ratio UWB antenna is still a difficulty. In addition, the previous two UWB antennas only use genetic algorithm in the design process, which has slow convergence speed and extremely consumes computing resources, so it needs to be improved combined with new methods. Because of its fast convergence speed, particle swarm optimization algorithm is often used in antenna optimization design in recent years [14, 15, 28]. There are also reports of binary particle swarm optimization algorithm [16–18], but this method is easy to premature. The combination of genetic algorithm and particle swarm optimization algorithm can globally improve the optimization efficiency. The hybrid algorithm of PSO and GA has also been reported in many literatures [19–21, 29] and used for array synthesis. Here, we combine the hybrid algorithm of PSO and GA with the grid binary coding method for the design of UWB antenna.

7.5.1 Automatic Design Process of Hybrid Evolutionary Algorithm

Similar to the previous section, the 2D metal area is designed to be 24 mm × 24 mm, divided into 10 × 10 grid cells, as shown in Fig. 7.20, adjacent cells extend slightly to avoid singularity connection and current continuity. The size of each unit is 2.8 mm × 2.8 mm, (w = h = 2.8 mm). The metal surface is discretized and represented by binary numbers "0" and "1".

In binary particle swarm optimization, position and velocity are two-dimensional binary solution space. In the initial population, the position and velocity of each particle are randomly generated. The population size is 100. The structure of the antenna is determined by the position of each particle in the particle swarm optimization algorithm. Hamming distance and sigmoid function are used to realize the transformation from real number solution space to binary number solution space. The sigmoid function is expressed as:

$$S(v_{mn,t}) = \frac{1}{1 + e^{-v_{mn,\,t}}} \tag{7.6}$$

$v_{mn,\,t}$ is the velocity of the m-th particle in the t-th iteration in N-dimensional solution space. The speed is updated by the following formula:

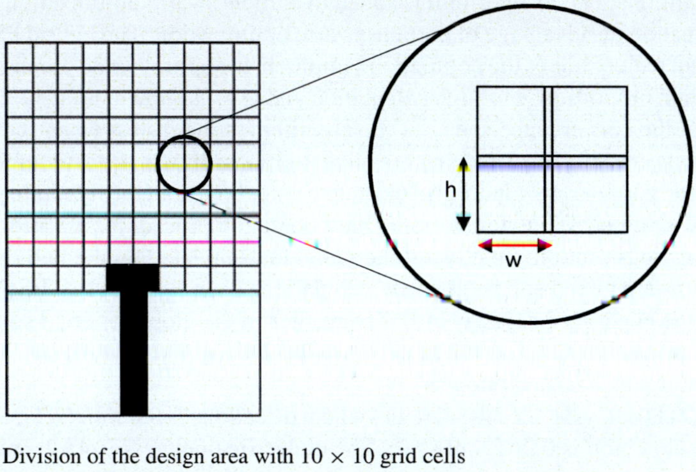

Fig. 7.20 Division of the design area with 10 × 10 grid cells

$$v_{mn,t} = v_{mn,t-1} + c_1\eta_1(p_{mn,t-1} - x_{mn,t-1}) + c_2\eta_2(g_{mn,t-1} - x_{mn,t-1}) \quad (7.7)$$

c_1 and c_2 are Hooke's constants. Generally, let c_1 and c_2 be equal to balance the local optimal and the global optimal. In addition, to randomize the population behavior, c_1 and c_2 are multiplied by a random factor η_1 and η_2 respectively, and η_1 and η_2 are random numbers uniformly distributed between (0, 1). $p_{mn,t-1}$ is the local optimal position of each particle, and records the best trajectory of each particle. $g_{mn,t-1}$ is the global optimal position of entire population, and it adjusts the global best trajectory of this population to accelerate the convergence speed of particle optimization.

The position vector of each particle is updated by the following formula:

$$x_{mn,t} = \begin{cases} 1, & r_{mn,t} < S(v_{mn,t}) \\ 0, & r_{mn,t} \geq S(v_{mn,t}) \end{cases} \quad (7.8)$$

$r_{mn,t}$ is a random number uniformly distributed between (0, 1). Compare it with $S(v_{mn,t})$, and the comparison result constitutes the position function of the m-th particle. In binary code of the position function, "1" means the metal grid of the antenna structure, and "0" means the non-metal grid. And the position and velocity of particles are updated continuly.

Here, FDTD is used to EM computation, and evaluate fitness function. The fitness function is same as described in the formula (7.15), is still expressed as:

$$fitness = \frac{1}{N}\sum_{i=1}^{N} S_{11}(i)$$

$$where\ S_{11}(dB) = \begin{cases} S_{11}(dB) + 15dB, & S_{11}(dB) \geq -15dB \\ 0, & S_{11}(dB) < -15dB \end{cases}$$

In addition, genetic algorithm is added to particle swarm optimization to improve the premature disadvantage of particle swarm optimization. The hybrid PSO and GA algorithm mainly mixes the population samples in the optimization process according to a certain proportion. Through calculation and test, as shown in Fig. 7.21, it can be seen that the convergence speed of the algorithm is the fastest when the number of individuals entering PSO and GA are allocated according to the ratio of 7:3. That is, 70% of the particles are randomly selected to enter PSO, and the remaining 30% enter 2-D GA for crossover and mutation. Other operation procedures are similar to those in the previous section. Finally, all the individuals updated by the two optimization algorithms are mixed and form a new population. Then evaluate the fitness function. Repeat the cycle successively until it converges to the global optimal solution.

The flow chart of using PSO and GA hybrid algorithm to design UWB antenna is shown in Fig. 7.22 as follows:

In FDTD analysis, the step size in x and y directions is 0.4 mm ($\Delta 1$), and the step size in z direction is 0.265 mm ($\Delta 2$). The dielectric thickness is $4\Delta 2$, and the width

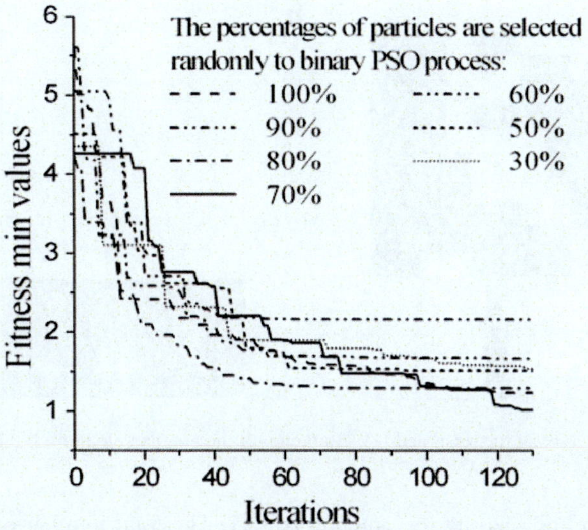

Fig. 7.21 The convergence rates from different percentages between PSO and GA

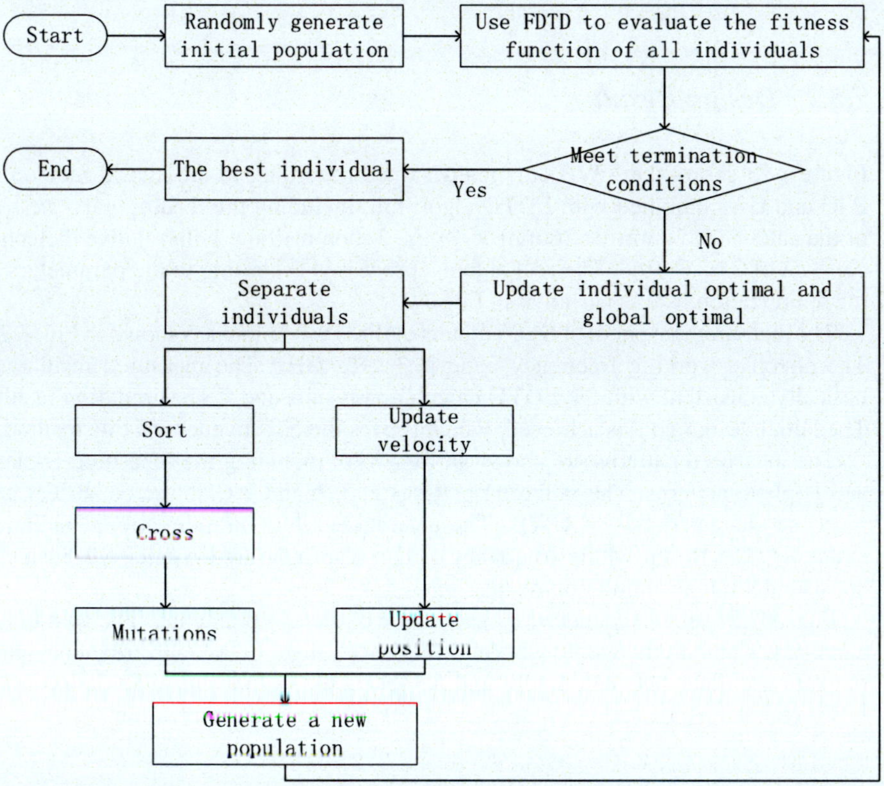

Fig. 7.22 The flow chart of the hybrid method of PSO and GA for the design of UWB antenna

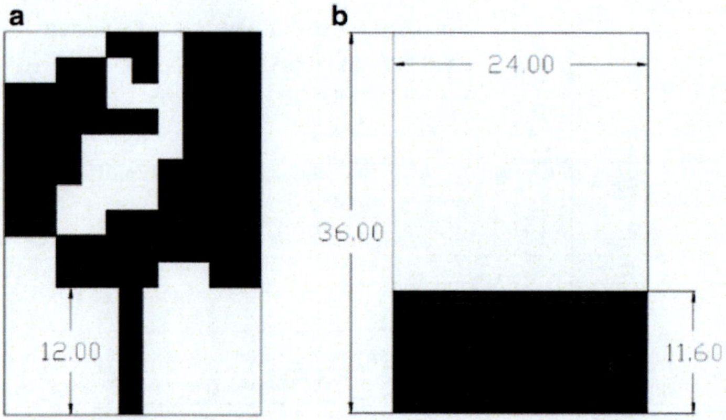

Fig. 7.23 Structure of the designed UWB antenna **a** the antenna element, **b** the ground

of microstrip line feeding the antenna is 8Δ1. The total number of grids is 100 × 100 × 37. Using unit Gaussian pulse excitation, the total time step of each particle is 6000Δt.

7.5.2 Design Result

Figure 7.23 shows the UWB antenna designed according to the hybrid method of PSO and GA combined with FDTD algorithm. Including the feeding part, the size of the antenna is 24 mm × 36 mm × 1 mm, Teflon medium with relative dielectric constant of 2.65 is used. The real antenna processed according to the parameters of the optimization results is shown in Fig. 7.24.

The measurement curve of S parameters of the UWB antenna is shown in Fig. 7.25. The effective working frequency band is 3.1–12 GHz. The measured results are basically consistent with the FDTD calculation results and CST simulation results. The initial design goal is achieved, which proves the effectiveness of this method.

The measured patterns are shown in Fig. 7.26, including the measured E-plane and H-plane patterns. The radiation patterns at 3, 6 and 9 GHz are consistent and meet the characteristics of UWB. The antenna exhibits omnidirectional radiation characteristics in the whole frequency range. The measured results are basically consistent with the simulation results.

Figure 7.27 shows the measured gain curve of the UWB antenna. The gain of the antenna is stable in the whole working frequency range. In the high frequency part, the gain oscillates between 4.4dbi and 5.3dbi. It shows that this antenna has good UWB characteristics and is suitable for UWB wireless communication system.

Fig. 7.24 Photograph of the designed UWB antenna

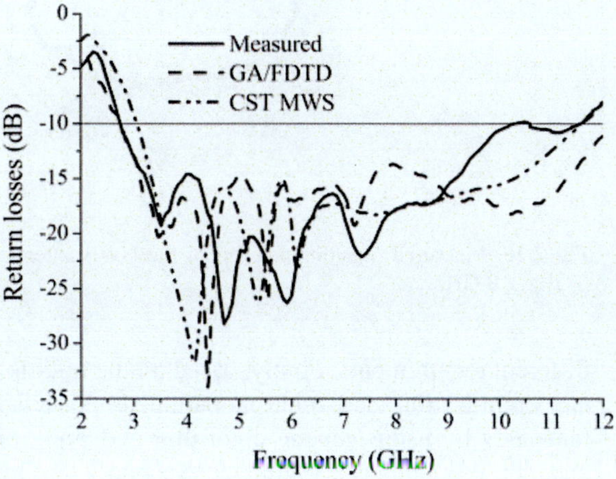

Fig. 7.25 Measured and simulated return losses for the UWB antenna

7.6 Summary

In this chapter, the automatic optimization design platform of UWB antenna by combining the optimization algorithm and FDTD electromagnetic calculation method is constructed, and optimizes several UWB antenna structures that meet

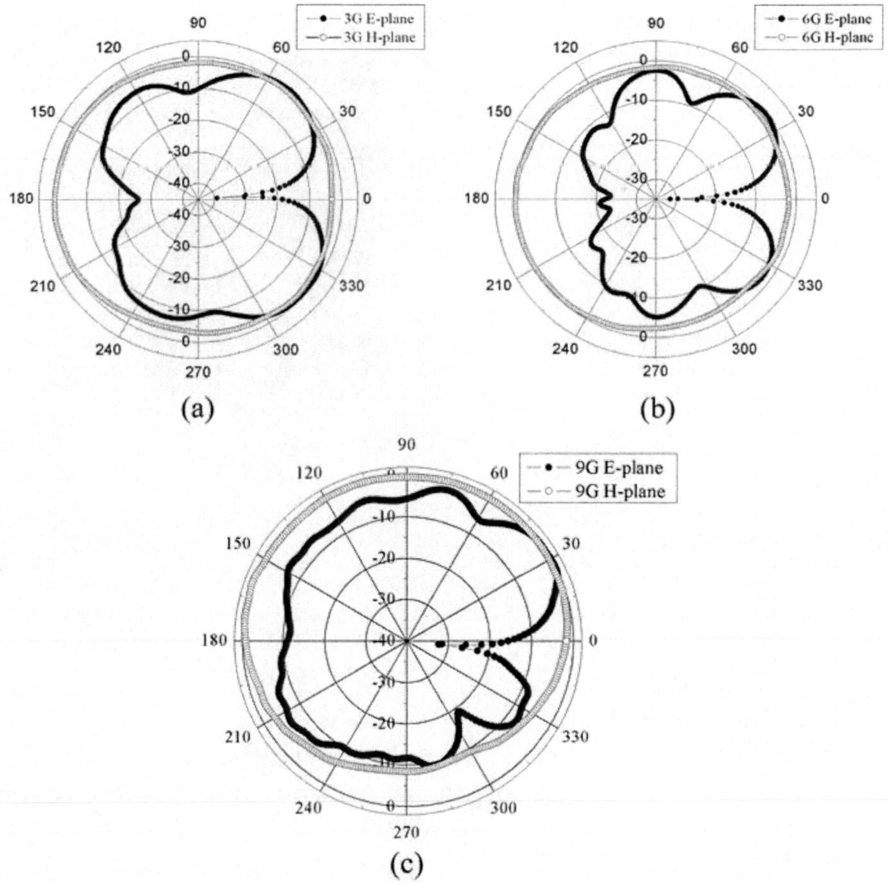

Fig. 7.26 Measured radiation patterns of the UWB antenna in different frequencies **a** 3 GHz, **b** 6 GHz, **c** 9 GHz

different requirements. Firstly, based on the traditional empirical model and parameter optimization idea, multiple parameters of UWB antenna are optimized simultaneously by using genetic algorithm and finite difference time domain method, and an UWB antenna is optimized. Secondly, two kinds of antenna structures, UWB antenna and band stop UWB antenna, are designed automatically by combining two-dimensional genetic algorithm and finite-difference time-domain method. Thirdly, in order to overcome the disadvantage of slow convergence speed, a miniaturized UWB antenna is designed by using the hybrid method of PSO and GA combined with FDTD algorithm. The optimization idea, platform and UWB antenna examples in this chapter prove that the fast automatic optimization platform is feasible.

Fig. 7.27 Measured gains of the UWB antenna in different frequencies

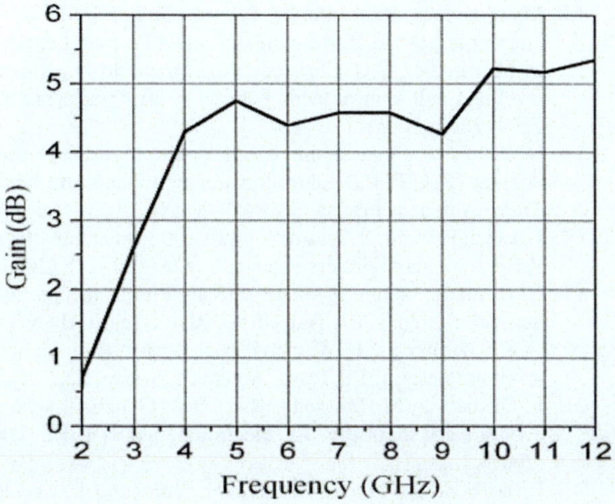

References

1. L. Wang, *Intelligent Optimization Algorithm and its Application* (Tsinghua University Press, 2001)
2. J. Kim, T. Yoon, J. Kim, Design of an ultra wide-band printed monopole antenna using FDTD and genetic algorithm. IEEE Microwave Wirel. Compon. Lett. **15**, 395–397 (2005)
3. N. Telzhensky, Y. Leviatan, Novel method of UWB antenna optimization for specified input signal forms by means of genetic algorithm. IEEE Trans. Antennas Propag. **54**, 2216–2225 (2006)
4. N. Telzhensky, Y. Leviatan, Planar differential elliptical UWB antenna optimization. IEEE Trans. Antennas Propag. **54**, 3400–3406 (2006)
5. L. Lizzi, F. Viani, R. Azaro, A. Massa, Optimization of a spline-shaped UWB antenna by PSO. Antennas and Wireless Propagation Letters. **6**, 182–185 (2007)
6. J. Martinez-Fernandez, J.M. Gil, J. Zapata, Ultrawideband optimized profile monopole antenna by means of simulated annealing algorithm and the finite element method. IEEE Trans. Antennas Propag. **55**, 1826–1832 (2007)
7. Z. Altman, New designs of ultra-broadband antennas using genetic algorithms, in *IEEE Antenna Propagation Society International Symposium*, Seattle, WA, pp 2054–2057 (1994)
8. D.S. Linden, E.E. Altshuler, Automating wire antenna design using genetic algorithms. Microw. J. **39**, 74–86 (1996)
9. M. John, M.J. Ammann, Wideband printed monopole design using a genetic algorithm. Antennas Wireless Propag. Lett. **6**, 447–449 (2007)
10. A.J. Kerkhoff, H. Ling, Design of a band-notched planar monopole antenna using genetic algorithm optimization. IEEE Trans. Antennas Propag. **55**(3), 604–610 (2007)
11. M. Ding, *The Researches on Broadband Antennas with Miniaturization*, PH.D. thesis (Shanghai Jiao Tong University, 2010)
12. B.-K. Yeo, Y. Lu, Array failure correction with a genetic algorithm. IEEE Trans. Antennas Propag. **47**, 823–828 (1999)
13. M. Ding, R. Jin, J. Geng, Optimal design of UWB antennas using a mixed model of 2-D GA and FDTD. Microw. Opt. Technol. Lett. **49**(12), 3177–3180 (2007)
14. N. Jin, Y. Rahmat-Samii, Advances in particle swarm optimization for antenna designs: real-number, binary, single-objective and multiobjective implementations. IEEE Trans. Antennas Propag. **15**, 556–567 (2007)

15. M. Benedetti, R. Azaro, D. Franceschini, PSO-based real-time control of planar uniform circular arrays. IEEE Antennas Wireless Propag. Lett. **5**, 545–548 (2006)
16. M.M. Khodier, C.G. Christodoulou, Linear array geometry synthesis with minimum sidelobe level and null control using particle swarm optimization. IEEE Trans. Antennas Propag. **53**, 2674–2679 (2005)
17. N. Jin, Y. Rahmat-Samii, Parallel particle swarm optimization and finite-difference time-domain (PSO/FDTD) algorithm for multiband and wide-band patch antenna designs. IEEE Trans. Antennas Propag. **53**, 3459–3468 (2005)
18. L. Lizzi, F. Viani, R. Azaro, A. Massa, Optimization of a spline-shaped UWB antenna by PSO. IEEE Antennas Wireless Propag. Lett. **6**, 182–185 (2007)
19. C.-F. Juang, A hybrid of genetic algorithm and particle swarm optimization for recurrent network design. IEEE Trans. Syst. Man Cybern. B Cybern. **34**, 997–1006 (2004)
20. D.W. Boeringer, D.H. Werner, Particle swarm optimization versus genetic algorithms for phased array synthesis. IEEE Trans. Antennas Propag. **52**, 771–779 (2004)
21. F. Grimaccia, M. Mussetta, R.E. Zich, Genetical swarm optimization: self-adaptive hybrid evolutionary algorithm for electromagnetics. IEEE Trans. Antennas Propag. **55**, 781–785 (2007)
22. F Yao, S. Zhong, *Study on UWB Printed Slot and Air Dielectric Antenna*. PH.D. thesis (Shanghai University, 2006)
23. D. Ge, Y. Yan, *The FDTD Method for EM Wave* (Xi'an Dian Zi Ke Ji University press, 2003)
24. D.E. Goldberg, *Genetie Algorithm Search, Optimization, Machine Learning* (Addison Wisley Publishing Company, 1989)
25. W. Tang, Y. Gu, *The Study on the GA Method and Application in the Structure Optimization*. PH.D. thesis (Dalian University of Technology, 2002)
26. M. Zhou, S. Sun, *The Principle and Application of GA* (National Defense Industry Press, 2002)
27. X. Wang, L. Cao, *Genetic Algorithm Theory, Application and Software Implementation* (Xi'an Jiao Tong University Press, 2002)
28. J. Robinson, Y. Rahmat-Samii, Particle swarm optimization in electromagnetics. IEEE Trans. Antennas Propag. **52**, 397–407 (2004)
29. J. Robinson, S. Sinton, Y. Rahmat-Samii, Particle swarm, genetic algorithm, and their hybrids: optimization of a profiled corrugated horn antenna. IEEE Antennas Propag. Soc. Int. Symp. Dig. **1**, 314–317 (2002)
30. M. Ding, R. Jin, J. Geng, Auto-design of band-notched UWB antennas using mixed model of 2D GA and FDTD. Electron. Lett. **44**(4), 257–258 (2008)

Chapter 8
Improved Hexagon Unit to Antenna Optimization by Binary Coding Method

8.1 Introduction

With the increasing requirements for wireless transmission rate and capacity, the development of broadband wireless communication has become very urgent. The corresponding broadband antenna has become one of the bottlenecks of wireless communication system. In particular, high-speed mobile systems urgently require low profile circular polarization antenna to be conformal with carrier, it can work in a very wide frequency band, and have high gain and stable pattern, so as to ensure the reliability of electromagnetic transmitting / receiving channel [1]. In addition, with the increasing number of electromagnetic systems, the electromagnetic environment is becoming more and more complex, and the circularly polarized CP wave has strong anti depolarization ability when transmitted in complex environment, which also urgently needs the circularly polarized antenna. As we all know, among many circularly polarized antennas, equiangular helical antenna is a typical UWB antenna with wide impedance bandwidth and good circularly polarized performance. However, its bidirectional radiation characteristics limit its application. Although it is suggested to use a microwave absorber to change its bidirectional radiation into directional radiation [2], it will lead to half of the energy loss, which is equivalent to half of the radiation efficiency. In addition, in the traditional directional antenna design, through λ/An ideal level plane is fixed at the distance of 4 to reflect electromagnetic waves. However, this method is sensitive to the distance of the reflecting surface and greatly limits the available bandwidth. Furthermore, PBG [3] or EBG [4] are used as reflecting surfaces respectively, but their axial ratio bandwidth satisfying circular polarization is relatively narrow.

In recent years, some people have introduced the optimization algorithm into antenna design to improve antenna performance. The semi-automatic antenna design in integrated learning particle swarm optimization (CLPSO) [5] provides a new idea

J. Geng and R. Jin, *Antenna Optimization and Design Based on Binary Coding*, Modern Antenna, https://doi.org/10.1007/978-981-16-7965-0_8

for parasitic shape optimization, and introduces two parasitic layers with rectangular holes [6]. However, the connection of two square element points in this design method is very possible. It can not explain the electromagnetic characteristics of this singular point connection in theory and engineering, which destroys the continuity of current, increases the uncertainty of electromagnetic calculation, and restricts the performance optimization of antenna. In this paper, a hexagonal element is proposed to generate a continuous parasitic layer discrete hexagonal mesh, so that the 2D adjacent elements are always connected line to line, which solves the discontinuity of singular point connection. By introducing the optimization algorithm and CST software, the hexagonal grid constituting the parasitic layer is optimized, and a low profile broadband antenna with bandwidth circular polarization and stable directional pattern is obtained.

Obviously, the binary coded antenna design method based on hexagonal grid expansion not only inherits the binary coded antenna based on square grid described in the previous chapter, but also improves the shortcomings of the previous methods. In particular, this method can easily realize rotational symmetry and is very suitable for the design of circular polarization antenna.

8.2 The Binary Coding Method and the Model and Geometry Structure of Antenna

8.2.1 Binary Coding with Hexagon Unit

The binary coding model based on the hexagon element is shown in Fig. 8.1.

The basic unit is a hexagon with side length *a*. These same hexagonal units are arranged outward in turn circle after circle. The 1st hexagonal unit is in the center, the continued six units are in the second circle, and are arranged counterclockwise around position 1 (No. 2, No. 3, …, No. 7), the 12 units in the third circle are arranged counterclockwise (No. 8, …, No. 19), and the 18 units in the fourth circle are arranged counterclockwise (No. 20, …, No. 37), The 24 units in the fifth circle are arranged in reverse clock order (No. 38, …, No. 61), and so on. The *n*th circle is arranged with (n − 1) * 6 hexagonal units and arranged in reverse clock order to form a grid area of hexagonal unit arrangement. The whole area is also a hexagonal area with a radius of about $\frac{(2n-1)}{2}\sqrt{3}a$, This is the antenna area where we want to optimize the design.

Figure 8.1b is some state of the binary coding, and the meshes are filled with '0' or '1'. When the meshes are filled with the corresponding materials, '1' means red color material, and '0' means the wathet material, and the corresponding map of the mesh with these binary codes is shown in Fig. 8.1c.

Similar as the above chapters, we can construct the real structure with two methods: filling spaces or digging holes. Filling spaces means the original space is empty with the virtual meshes, and every mesh is filled with copper or air. Digging

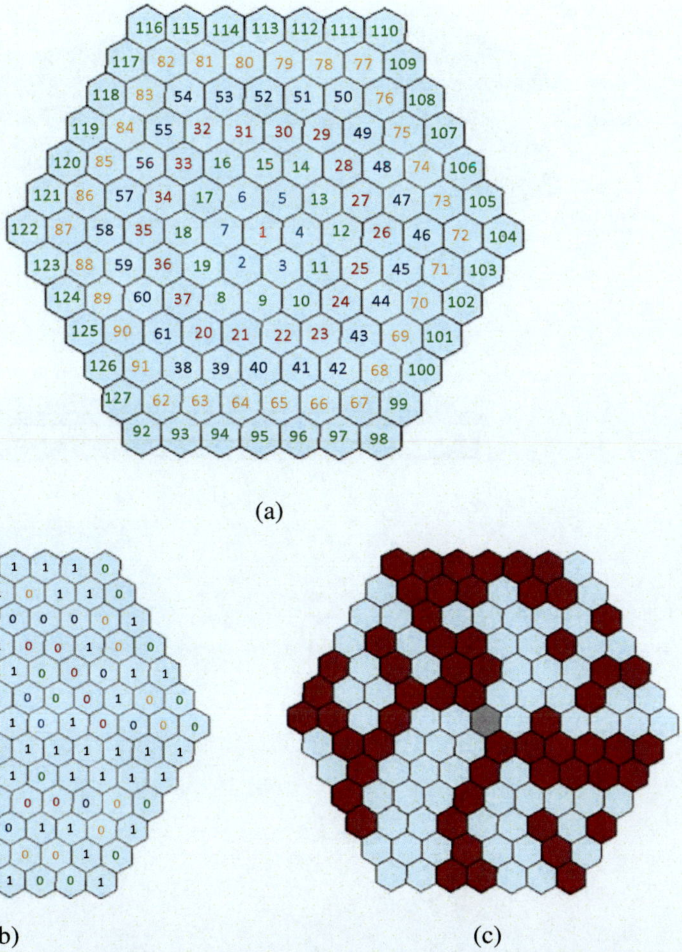

Fig. 8.1 Binary code to the 2D region based on the hexagon unit. **a** coding mode and sequence, **b** binary coding sample and **c** the corresponding patch with metal and space

holes means the original space is a hexagon area of metal with virtual meshes, and every mesh is digged a small hexagon hole or the metal is kept.

8.2.2 Antenna Prototype and Optimization

8.2.2.1 Antenna Structure

Figure 8.2 presents the structure of the antenna, which consists of antenna layer, two parasitic layers and ground layer. Three substrate layers used for the design is

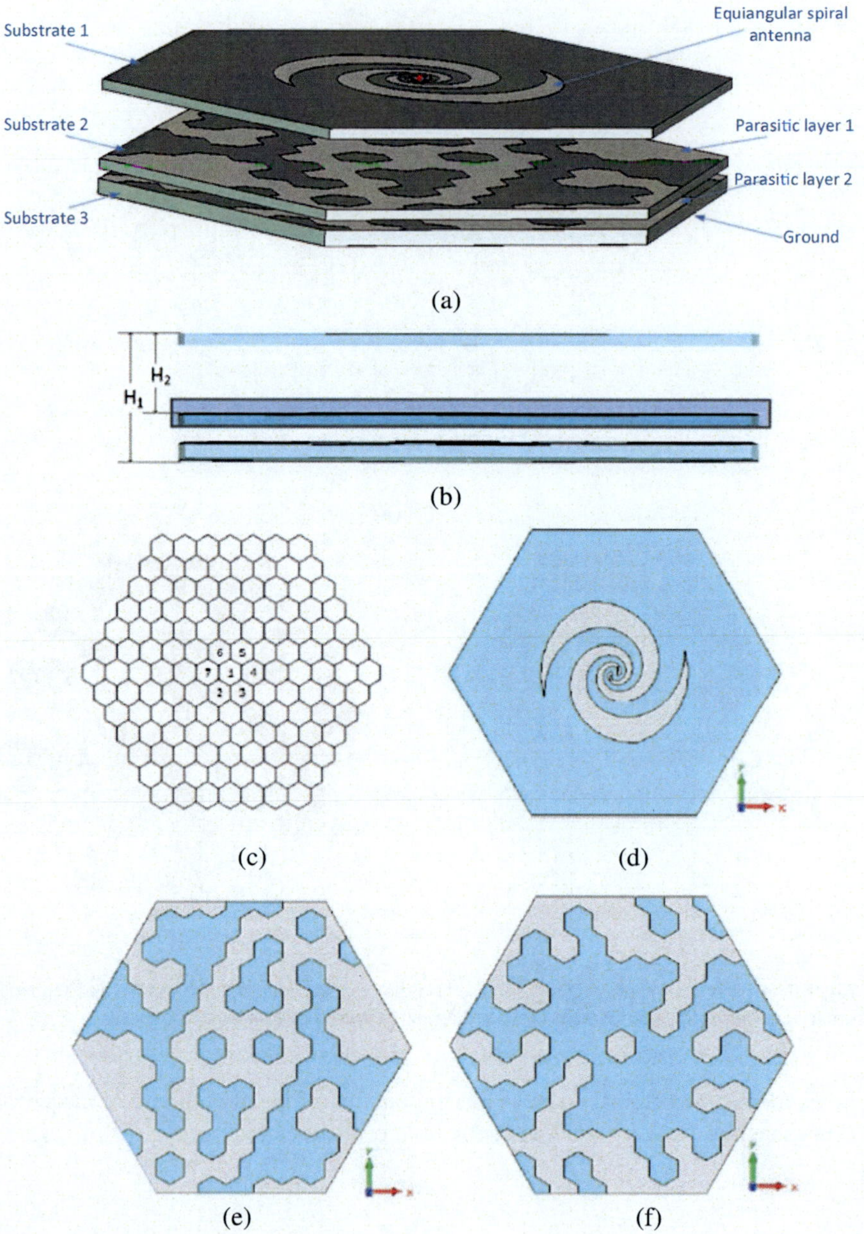

Fig. 8.2 Structure of proposed antenna. **a** 3D perspective view, **b** Side view, **c** Hexagon grids model, (d) Equiangular spiral antenna layer, **e** Parasitic layer 1, **f** Parasitic layer 2. Figure reproduced with permission from Ref. [7], © 2015 IEEE

Taconic TLT-6 with permittivity 2.65. The height of the antenna is H1 = 23.7 mm, and the distance between the antenna and parasitic layer is H2 = 20.7 mm. The top layer is the two arm equiangular spiral antenna with the 1.57 turns. In this model, two arm equiangular spiral antenna has a wide impedance band.

To improve the bandwidth of the antenna stable unidirectional pattern and axial ratio, parasitic layers are necessary. In [6], rectangular grids design based on semiautomatic optimization [5] is adopted. However, point-to point connection will occur in this model. In fact, we usually try to replace corner curve by smooth curve and point-to-point connection by line-to-line connection when we design antenna. Thus, we adopt regular hexagon as the basic element. Figure 8.2c shows the regular hexagon grids. Taking grid 1 as an example, there are grids 2–7 next to grid 1, and they are connected by 6 edges of grid 1 separately. Therefore, there are only slots or line-to-line connections in Fig. 8.2e, f if hexagon elements are adopted. Moreover, the hexagon has a better rotational symmetric performance, which is suitable to two arm equiangular spiral antenna. After PSO integrated optimization of the grids in CST, we achieve a low profile CP antenna.

8.2.2.2 Fitness Function

For the target antenna, on the one hand, we hope to have sufficient bandwidth, on the other hand, we hope to achieve low profile directional radiation. The desired bandwidth is given artificially, and the low profile is often a predetermined height, which can be realized through the range of height optimization. Therefore, the optimized fitness function can be set by the requirements for S11 of the desired frequency band and the gain in the normal direction of the antenna plane. S11 set the target to −15 dB, and for broadband antenna, set the gain target to 5 dB.

$$fitness = \frac{1}{N}\sum_{i=1}^{N}\{|S_{11}(f_i) - (-15)| + |G_{\theta=0}(f_i) - 5|$$
$$f_i = \{f_i : f_L \le f_i \le f_H, i = 1, \ldots, N\} \quad (8.1)$$

If we want the good circular polarization performance, the AR should be less 3 dB in the target band, and the fitness can be revised as

$$fitness = \frac{1}{N}\sum_{i=1}^{N}\{|S_{11}(f_i) - (-15)| + |G_{\theta=0}(f_i) - 5| + |\mathrm{AR}(f_i) - 3|$$
$$f_i = \{f_i : f_L \le f_i \le f_H, i = 1, \ldots, N\} \quad (8.2)$$

8.3 Result Analysis

The simulated S_{11} versus frequency is shown in Fig. 8.3, which is fed by discrete port. It is obvious that proposed antenna has a −10 dB impedance bandwidth of 229%, ranging from 2.1 to 7.18 GHz. Figure 8.4 further shows the simulated far field patterns at 2.9 GHz. From simulated results in Fig. 8.5, the proposed antenna has a broad stable unidirectional radiation bandwidth from 2.2 to 5.4 GHz. And Fig. 8.6 plots the axial ratio versus frequency. The axial ratio less than 3 dB extends from 2.4 to 3.6 GHz. Comparing with results of adopting rectangular model, there are no frequency points at which the axial ratio is bigger than 3 dB. In Fig. 8.7, the simulated realized gain curve over frequency is shown, and the realized gain is 6–10 dBi in the available band.

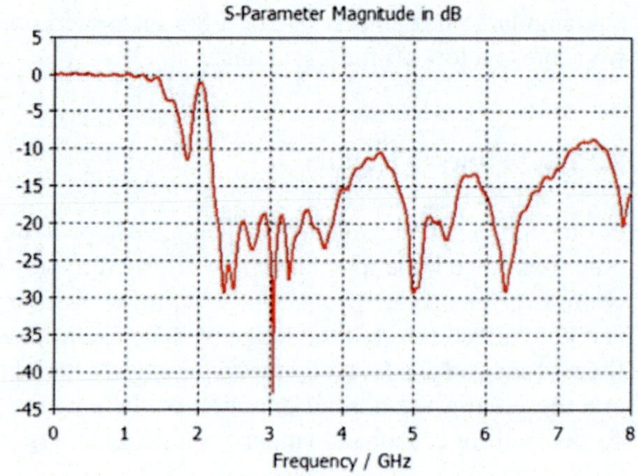

Fig. 8.3 Simulated S_{11}. Figure reproduced with permission from Ref. [7], © 2015 IEEE

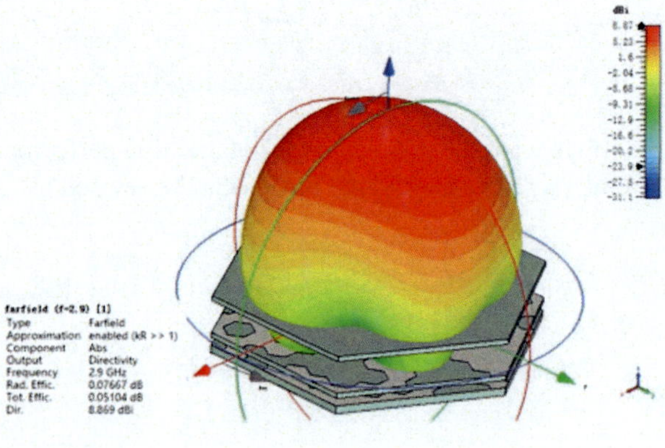

Fig. 8.4 Simulated far field pattern at 2.9 GHz

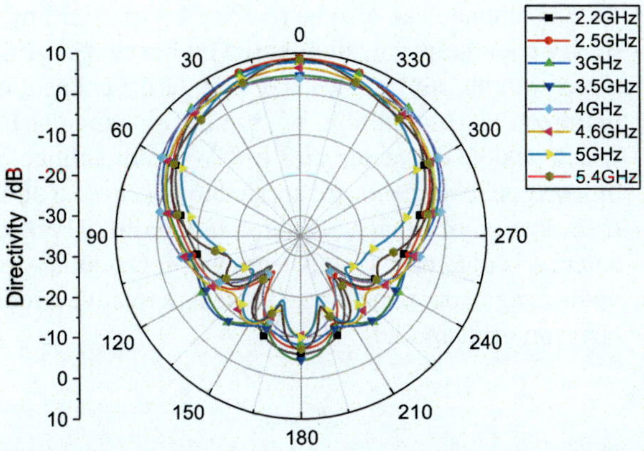

Fig. 8.5 The simulated far fields at 2.2, 2.5, 3, 3.5, 4, 4.6, 5 and 5.4 GHz, they all are directional upward

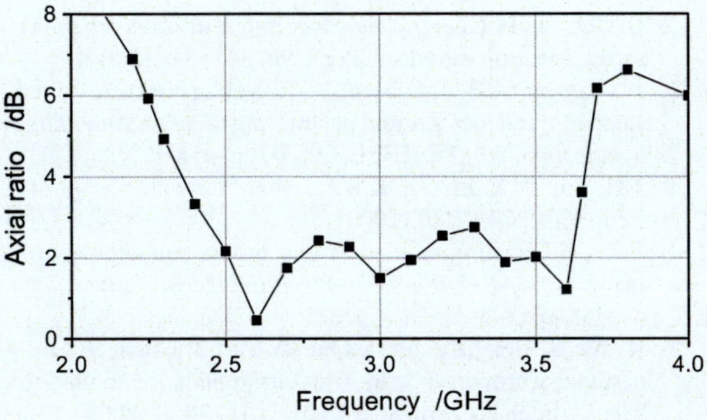

Fig. 8.6 Simulated axial ratio

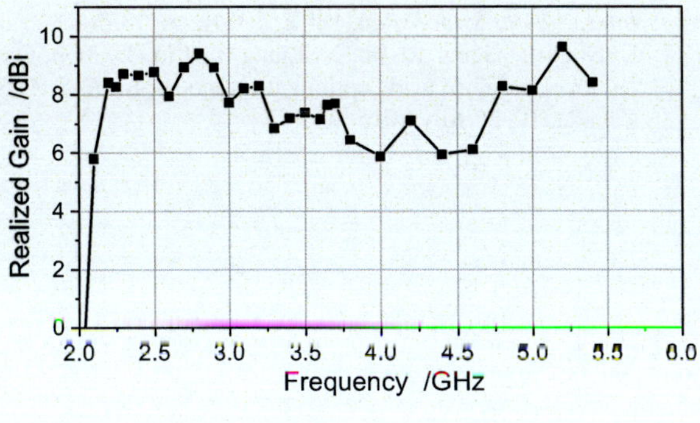

Fig. 8.7 The simulated realized gain

8.4 Summary

In this chapter, the hexagonal element is used as the basic element to construct the grid, and the binary codes of the hexagonal grids are constructed by circle and circle arrangement. Similar to the previous chapters, the conversion from binary code to antenna can also be realized by space filling or hole digging. Obviously, the binary coding optimization based on hexagonal grid makes the structure of antenna more smooth, and avoids the singularity connection problem in rectangular grid optimization. In addition, we use the equiangular helix as the basic radiation unit, which has the characteristics of UWB bidirectional radiation. Then, the grid model of hexagonal elements are applied to construct a double-layer reflector. Through the optimization of binary coding, a low profile UWB directional circular polarization antenna is obtained. Comparing with rectangular grids parasitic layers, the antenna with hexagon parasitic layers presents a relative wider axial ratio band, which verifies hexagon grids model.

References

1. D. Chen, C.H. Cheng, A novel compact ultra-wideband (UWB) wideslot antenna with via holes. Progr. Electromagn. Res. PIER **94**, 343–349 (2009)
2. J. Thaysen, K.B. Jakobsen, H.-R. Lenler-Eriksen, Wideband cavity backed spiral antenna for stepped frequency ground penetrating radar, in *Antennas and Propagation Society International Symposium*, vol. 1B (IEEE, 2005), pp. 418–421
3. T.H. Liu, W.X. Zhang, K.F.A. Tsang, Low profile spiral antenna with PBG substrate. Electron. Lett. **36**(9), 779–780 (2000)
4. Z. Li, G.M. Wang, Y. Cao, A low-profile equiangular spiral antenna using a novel EBG ground plane, in *ISAPE' 06. 7th International Symposium on Antennas, Propagation & EM Theory* (2006), pp. 1–3
5. H. Wu, J. Geng, R. Jin, J. Qiu, W. Liu, J. Chen, S. Liu, An improved comprehensive learning particle swarm optimization and its application to the semi-automatic design of antennas. IEEE Trans. Antennas Propag. **57**(10), 3018–3028 (2009)
6. X. Liu, J.P. Geng, R.H. Jin, X.L. Liang, K. Wang, Study the low profile CP antenna by combiningdiscrete grid model and parameter optimization, in *Antennas and Propagation Society International Symposium* (IEEE, 2014), pp. 1335–1336
7. L. Duan, J. Geng, R. Jin, X. Liang, L. Liu, J. Chen, C. He, A low profile CP antenna based on novel hexagon grids optimization model, in *IEEE APS 2015*, Vancouver, British Columbia, Canada, 19–25 July 2015

Chapter 9
Antenna Design Based on Optimization of Linear Motion Trajectory

A novel method to design patch antenna with computer-aid is presented in this chapter. Smooth curve is used to generate the border of the patch antenna, and particle swarm optimization (PSO) algorithm is implemented to control the shape of the patch, which leads the result less dependent on designer's experience and more reasonable. Finite Difference Time Domain (FDTD) algorithm is used for electromagnetic simulation. The optimization of process is conducted on the parallel cluster to reduce computational time. To verify the correctness of this method, an antenna working on WLAN band has been optimized and fabricated. The experiment result is showed in the end.

Antenna is always designed upon the experience of the designer, which is an onerous and time-consuming work. Some parameters of the antenna, such as directivity and S-parameter of monopole antenna, are not harmony, which leads great difficulty for the designer to modify the structure for proper working conditions. Till now, most patch antenna structure inherits previous successful case. Some kinds of unknown structure may perform better or even be extrapolated as a pattern for antenna design.

Recently, as the huge advancement of computer speed, some kinds of antenna optimization methods have been presented. But the optimized structure is often discontinuous. There are some flaws that grid meshing optimization cannot overcome in antenna optimization [1]. The electric characteristic of two diagonal-connected blocks is unknown and also leads to great difficulty in fabricating.

Therefore, a novel way is proposed in this chapter to implement computer-aided antenna design with smooth border. In this chapter, the configuration of the particle swarm optimization algorithm is presented. And a continuous and smooth structure generating method is described. In the end, the optimized result and experiment data are analyzed.

J. Geng and R. Jin, *Antenna Optimization and Design Based on Binary Coding*, Modern Antenna, https://doi.org/10.1007/978-981-16-7965-0_9

9.1 Application of Bezier Curve in Antenna Design

In order to apply the Bezier curve to the antenna design, this section proposes the following design methods, as shown in Fig. 9.1. First, the entire plane is divided into two parts: Control Area and Object Area. Among them, the Control Area contains the target area. The control area is an area where the control points are freely distributed, and the target area limits the range of the generated continuous smooth curve, that is, the curve cannot exceed the range of the target area. In actual manufacturing, the target area is also the upper part of the patch antenna. The points in the control area are divided into two categories: Fixed Points and Control Points. The fixed point is the start and end point of a continuous smooth curve($\mathbf{P}_0$ and $\mathbf{P}_n$),Generally, the patch antenna adopts the side-fed feed point mode, so the fixed point is the position connected with the feeder, and different spacing is generally used according to the different impedance requirements (General impedance requirement is 50 Ω). As the name implies, control points ($\mathbf{P}_1 \sim \mathbf{P_{n-1}}$) are used to control the distribution of the entire continuous curve. The PSO optimization algorithm controls the position of these control points on the plane. In order to speed up the optimization process, we slightly adjust the position of the fixed point according to the different control points to meet the common requirements for the connection of the antenna feeder and the radiating element. If suppose $\mathbf{P}_{0x} < \mathbf{P}_{\mathbf{n}x}$, when the control point meets the following conditions:

$$\mathbf{P}_{1x} > \mathbf{P}_{\mathbf{n}} - 1x \tag{9.1}$$

Then swap the positions of $\mathbf{P}_0$ and $\mathbf{P}_n$.This ensures the connectivity between the feeder and the antenna unit.

Fig. 9.1 Specific application of Bezier curve in antenna design

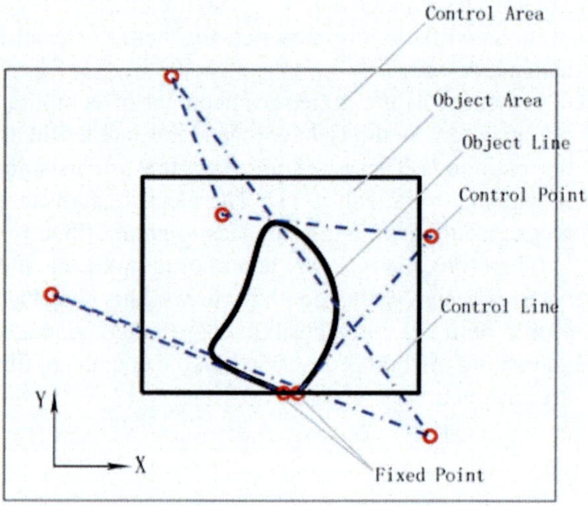

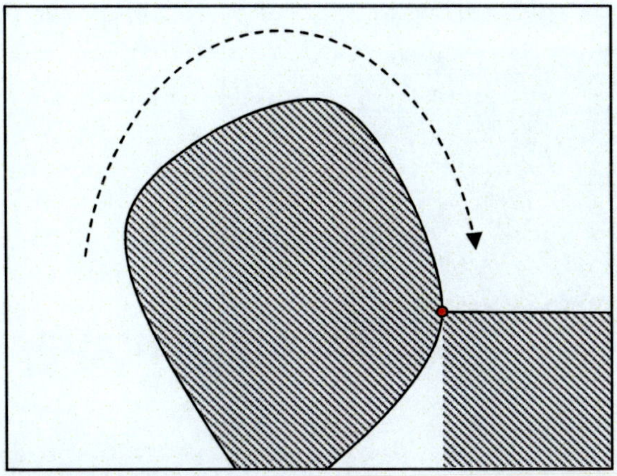

Fig. 9.2 Example of curve filling step

9.2 Curve Filling

Before using the Bezier curve formula to generate a continuous smooth curve to describe the boundary, how to determine which part of the area surrounded by the curve is metal or not is a problem to be solved, especially when the curves cross each other. Therefore, this section proposes a simple and effective method of curve filling.

As shown in Fig. 9.2, the entire curve filling method is divided into three steps:

(1) The state of each grid point on the plane is divided into two types: filled and unfilled, which can be represented by 1 and 0.
(2) Perform the following operations for each point on the curve: Invert the state of all points to the right of the point.
(3) Traverse all points on the curve.

In Fig. 9.3, the right is a randomly generated smooth curve, and the picture on the left is the result after filling.

Judging from the filling effect, this method is proved to be fast and correct.

9.3 Optimal Design and Smooth Curve

As the electromagnetic field is smooth, the antenna with smooth structure will match the field well and may lead to improvement in working performance. In this section, particle swarm optimization algorithm [2–5] is used to optimize the smooth border of the patch antenna. The flow chart of particle swarm optimization is shown in Fig. 9.4. The update equation is:

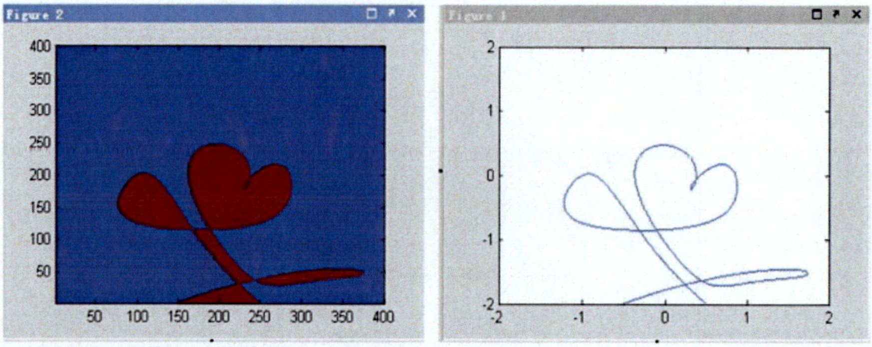

Fig. 9.3 Curve filling example

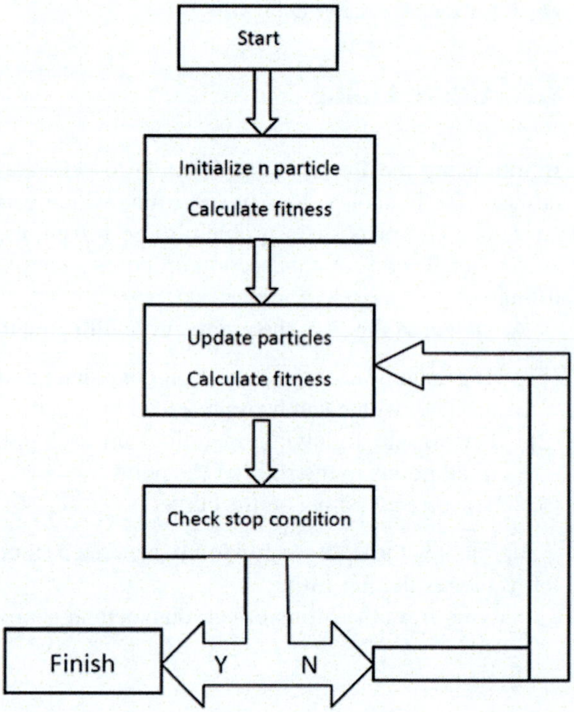

Fig. 9.4 Flow chart of particle swarm optimization

$$V_i^{k+1} = \omega V_i^k + c_1 \xi (P_i^k - X_i^k) + c_2 \eta (p_g^k - X_i^k),$$
$$X_{id}^{k+1} = X_i^k + r V_i^{k+1} \tag{9.2}$$

in which, $x_i(x_{i1}, x_{i2}, \ldots, x_{iD})$ is the position vector of particle i, $V_i(V_{i1}, V_{i2}, \ldots, V_{iD})$ is the velocity vector of particle i, $P_i(P_{i1}, P_{i2}, \ldots, P_{iD})$

is the local best vector of particle *i*, $P_g(P_{g1}, P_{g2}, \ldots, P_{gD})$ is the global best vector, note that there is only one P_g, ω is inertia factor, which is used to keep original velocity, c_1 is local best factor, which is used to keep track with the local best vector, c_2 is global best factor, which is used to keep track with the global best vector. ξ and η are random numbers in [0, 1], *r* is constraint factor, which is used to update the position.

In order to generate smooth curve of any arbitrary shape, following equation [6] is used.

$$\mathbf{P}(t) = \sum_{i=0}^{n} C_n^i (1-t)^{n-i} t^i \mathbf{P_i} \tag{9.3}$$

t is a real number and changes from 0 to 1.
P is a set of points to describe the Smooth curve.
P_i is the control point, in which changes from 0 to *n*.

9.4 Optimization Results and Analysis

In order to verify the feasibility of the continuous smooth boundary and the performance of the entire optimization platform, this section finally targeted the 3-band patch antenna for optimization. The specific settings are as follows:

Table 9.1 List of optimization goals

Antenna type	Patch antenna
Medium parameter (ε_r)	2.65
Medium thickness	1 mm
Metal thickness	0.05 mm
Input resistance	50Ω
Antenna size	70 mm × 60 mm × 1 mm
Target area	70 mm × 60 mm × 1 mm
Feeder length	10 mm
Working frequency	*WLAN*(2.45 GHz, 5.25 GHz, 5.8 GHz), Tri-band
Optimization requirements	Working frequency S_{11} < 15 dB
Number of curve control points	15

9.4.1 Optimization Goal

As shown in Table 9.1, the desired antenna can work on 3 frequencies, that is, this is an antenna optimization problem with 3 objectives.

9.4.2 Fitness Function

Similarly, the fitness function of the multiband antenna can be constructed as following.

$$F(\mathrm{dB}) = \sum \left\{ S_{11}(f_i) + \frac{1}{5}[S_{11}(f_i + \Delta) + S_{11}(f_i - \Delta)] \right\} \tag{9.4}$$

Among them, f_i is a set of required operating frequency points, here are 2.45, 5.25, and 5.8 GHz, Δ is bandwidth requirement, generally 50 MHz.

9.4.3 The Optimization Results

As shown in Fig. 9.5, the dark part is metal.

From the optimization results, the antenna has a smooth boundary, which meets the initial design requirements.

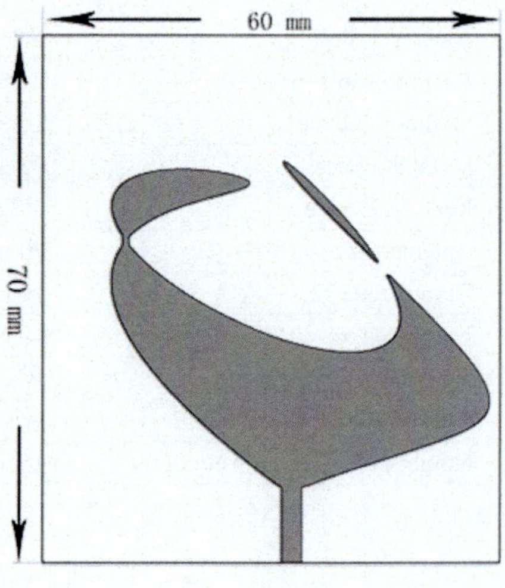

Fig. 9.5 Optimization Results. Figure reproduced with permission from Ref. [7], © 2010 IEEE

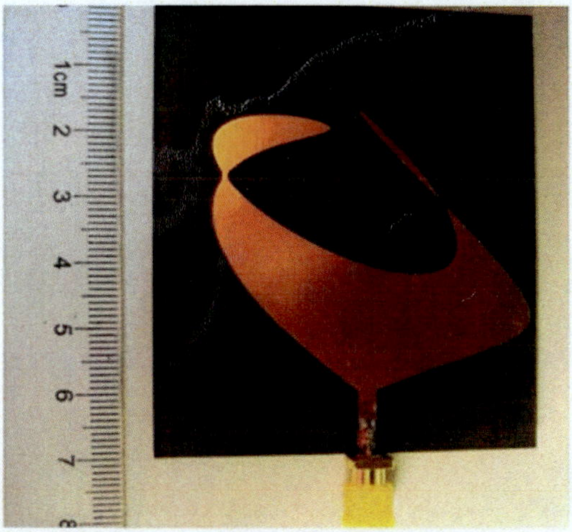

Fig. 9.6 The fabricated prototype antenna. Figure reproduced with permission from Ref. [7], © 2010 IEEE

9.4.4 Physical Photo

The photo of real object is shown in Fig. 9.6.

9.4.5 Simulation and Measurement Results

The return loss parameters of the antenna are shown in Fig. 9.7. The blue solid line is the simulation result, and the red dashed line is the measurement result. The antenna pattern characteristics are shown in Fig. 9.8. The blue solid line is the simulation result, and the red dashed line is the measurement result.

The theoretical value and the measured value are basically consistent, but there is still a certain error. After further simulation verification, the reasons for the error may be as follows:

(1) Error is inevitable.
(2) The dielectric constant of the dielectric plate is lower than the design value (which is further verified by simulation).
(3) The antenna feed connector (SMA head) is of poor quality.
(4) Large errors occur when the continuous smooth structure is processed in the simulation.
(5) Processing error.

However, the overall trend is right, and the possibility of the second and fourth reasons is very high. Further improvements in simulation software and processing level are needed to solve this deviation. But it has been proven to work.

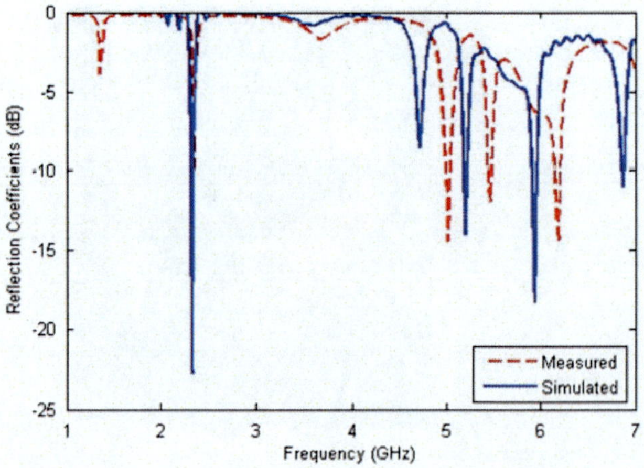

Fig. 9.7 The measured and simulated S_{11} values of optimized antenna parameters. Figure reproduced with permission from Ref. [7], © 2010 IEEE

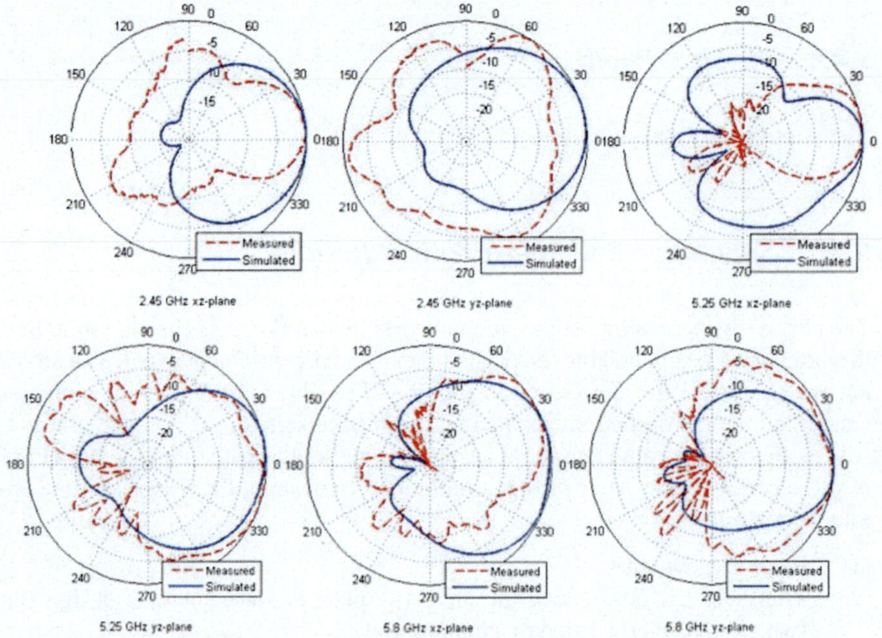

Fig. 9.8 Antenna pattern characteristics at different frequencies and directions. Figure reproduced with permission from Ref. [7], © 2010 IEEE

9.5 Summary

This chapter mainly studies the multi-objective antenna optimization design method, builds a parallel optimization design platform for antennas based on a hybrid platform, and proposes the optimization idea of smooth boundaries. Finally, a tri-band antenna is optimized, then tested and verified.

References

1. M. Ding, R. Jin, J. Geng, Optimal design of ultra wideband antenna using a mixed model of 2-D genetic algorithm and finite-difference time-domain. Microw. Opt. Technol. Lett. **49**(12), 3177–3180
2. R. Eberhart, P. Simpson, R. Dobbins, *Computational Intelligence PC Tools* (Academic Press, Inc., 1996), pp. 212–223
3. J. Kennedy, R. Eberhart, Y. Shi, *Swarm Intelligence* (Academic Press, Inc., 2001)
4. B. Birge, PSOt, A particle swarm optimization toolbox for matlab, in *IEEE Swarm Intelligence Symposium Proceedings*, 24–26 Apr 2003
5. H. Wu, J. Geng, R Jin, J Qiu, W. Liu, J. Chen, S. Liu, An improved comprehensive learning particle swarm optimization and its application to the semi-automatic design of antennas. IEEE Trans. Antennas Propag. **57**(10), 3018–3028 (2009)
6. T.W. Sederberg, S.R. Parry, Free-form deformation of solid geometric models, in *ACM Siggraph Computer Graphics* (1986)
7. B. Zhou, J. Geng, R. Jin, X. Liang, S. Ye, The optimal design of patch antennas with smooth border, in International Workshop on Antenna Technology (iWAT), *1-3 March 2010.*